Generis
PUBLISHING

Green Innovation Frontiers

A Journey Through Senegal's Sustainable Development in Construction, Energy, and Agriculture

Adama Doudou Diop

Title: Green Innovation Frontiers

A Journey Through Senegal's Sustainable Development in Construction, Energy, and Agriculture

ISBN: 979-8-89966-239-3

Author: Adama Doudou Diop

Cover image: https://unsplash.com/

Publisher: Generis Publishing
Online orders: www.generis-publishing.com
Contact email: info@generis-publishing.com

Table of Contents

Acronyms and Abbreviations

AFD: French Development Agency
ANER: National Renewable Energy Agency
ASN: Senegalese Standards Agency
ANSD: National Agency for Statistics and Demography
AVN: Nubian Vault Association
AVF-Sénégal: Fertile Green Africa Senegal
BFS: Bureau for Food Security (USAID)
BTP: Building and Public Works
BRADES: Research/Action Office for Solidarity-Based Development
CDH: Horticulture Development Center
ECOWAS (CEDEAO): Economic Community of West African States
CEEDD: Listening and Support Center for Sustainable Development
CFD: Training and Demonstration Center
CGAAER: General Council for Food, Agriculture and Rural Areas (France)
CIAT: International Center for Tropical Agriculture
CPC: Community Production Center
CNCR: National Council for Rural Dialogue
CSE: Ecological Monitoring Center
DEEC: Directorate of Environment and Classified Establishments
DIEM: Directorate of Infrastructure, Equipment, and Maintenance
ECREEE: ECOWAS Centre for Renewable Energy and Energy Efficiency
RE (EnR): Renewable Energy
FAO: Food and Agriculture Organization of the United Nations
IPCC (GIEC): Intergovernmental Panel on Climate Change
GWh: Gigawatt-hour
GOANA: Grand Agricultural Offensive for Food and Abundance
HLM: Low-Income Housing
LPDSE: Energy Sector Development Policy Letter
LP: LePartenariat
OAPI: African Intellectual Property Organization
NGO (ONG): Non-Governmental Organization
OMVG: Organization for the Development of the Gambia River
OMVS: Organization for the Development of the Senegal River
MAER: Ministry of Agriculture and Rural Equipment
MEDD: Ministry of Environment and Sustainable Development

MEDER: Ministry of Energy and Renewable Energy Development
MEPN: Ministry of Environment and Nature Protection (former MEDER)
MRUHCV: Ministry of Urban Renewal, Housing and Living Environment
MW: Megawatt
PANERs: National Renewable Energy Action Plan of Senegal
PERACOD: Program for the Promotion of Renewable Energies, Rural Electrification, and Sustainable Supply of Domestic Fuels
PETN: Taiba N'diaye Wind Farm
PDIDAS: Inclusive and Sustainable Agribusiness Development Project in Senegal
PRACAS: Program for Accelerating the Pace of Senegalese Agriculture
PRODERE: Renewable Energy Development Program
PSE: Emerging Senegal Plan
RFI: Radio France Internationale
RGPH: General Population and Housing Census
RGPHAE: General Census of Population, Housing, Agriculture, and Livestock
SABER: African Company of Biofuels and Renewable Energies
SICAP SA: Former Cape Verde Real Estate Company
SNHLM: National Company for Low-Income Housing
SOGAS: General Abattoir Company of Senegal
SDE: Senegalese Water Company
ICT (TIC): Information and Communication Technologies
TER: Regional Express Train
UADB: Alioune Diop University of Bambey
UEMOA: West African Economic and Monetary Union
UGB: Gaston Berger University of Saint-Louis
UNESCO: United Nations Educational, Scientific and Cultural Organization
USAID: United States Agency for International Development
UV: UltravioletVN: Nubian Vault

Abstract

In the current contexts of climate change, environmental degradation, and sustainable development, several actions considered ecological and innovative are being promoted and spread across the world. However, questions still arise about their relevance and effectiveness, particularly in countries in the Global South. The goal of this study is to analyze the challenges of some of these innovations in Senegal through a geographical approach in the sectors of eco-construction, alternative energy, and sustainable agriculture. The methodology involved selecting two innovations from each sector, then conducting comparative analyses, direct observations, interviews, and, in one case, a focus group. The six innovations studied are: Nubian vault (VN), compressed earth blocks (BTC), wind farm, biochar, urban micro-gardening, and biofertilizers. The results and their discussions showed that these ecological innovations, in the context of Senegal, are interesting, but the challenges are numerous. They have not yet reached a level where they can effectively replace what is already being done in the studied sectors (except possibly for biofertilizers, if their production was done by the beneficiaries themselves). Considering these parameters can open up several perspectives for adaptation.

Keywords: Ecological innovations; Eco-construction; Alternative energies; Sustainable agriculture; Sénégal

Acknowledgments

We extend our deepest gratitude to all those who, directly or indirectly, contributed to the successful completion of this work. Our thoughts are with the late Professor Mouhamadou Mawloud Diakhaté, with whom we initiated this research in 2019 and continued extensively during our Master's studies. Under his guidance, we learned the values of self-determination and academic freedom. May his soul rest in peace.

Our heartfelt thanks also go to the individuals and institutions who facilitated the acquisition of field data:

- The City of Dakar, through all the stakeholders involved in the micro-garden initiatives;
- Mr. Malick Sy of the NGO Le Partenariat in Podor;
- Mr. Hubert Dieme and Ms. Pauline Georges of the NGO Nebeday in Toubacouta;
- Mr. Oumar Wane of Biotech Services Senegal;
- Mr. Amadou Sow of the Taiba Ndiaye Wind Farm / Lekela;
- Mr. Chris Ford of Lekela, London;
- Mr. Ablaye Fall of the Taiba Ndiaye Municipal Council;
- Mr. Loïc and Mr. Amadou Diallo of Village Pilote in Dakar;
- Mr. Théodore Ndiaye of Elementerre in Gandigal (Mbour).

To all, we express our sincere appreciation for your invaluable contributions.

General Introduction

Can Innovation and New Technologies Solve the Environmental Crisis? What if Human Intelligence Could Lead Us Out of the Ecological Emergency?

The ecological crisis today represents a major threat to the sustainability of our societies. Climate change, biodiversity loss, ocean acidification, air pollution, and the depletion of natural resources are all deeply challenging our economic, agricultural, and social development models. These are no longer distant or abstract threats—their impacts are already visible and measurable. For instance, agricultural yields are declining due to global warming, environmental and climate-related migrations are increasing, and ecosystems are becoming increasingly fragile as biodiversity collapses.

In light of this alarming situation, a fundamental question arises: What can we do? More specifically, can technological innovation and human intelligence enable us to effectively confront this global crisis?[1]

Sustainable development and ecological initiatives have become nearly unavoidable in today's international relations. This form of development, which aims to be more equitable and environmentally conscious, seeks to "meet the needs of the present without compromising the ability of future generations to meet their own needs" (Brundtland Report, 1987). As Brunet et al. (1997) put it, "The damages caused by progress are perceived as threats to the future of humanity and a nuisance to its present, calling for action and the abandonment of certain operations."

It is in this context that ecological innovations and terms such as "organic," "eco," "green," or "environmentally friendly" have acquired particularly positive connotations, occupying a significant place in scientific, media, academic, and even political discourse. In the Global South, this trend has led to the adoption and funding of numerous initiatives by NGOs, international organizations, development partners, and public authorities.

However, several questions arise given the persistence of many environmental and developmental challenges: What are the constraints facing such approaches, which are nonetheless widely regarded as important and necessary by public opinion? What challenges remain to be addressed?

To answer these questions, this work focuses its analysis on Senegal, through the lens of three key sectors:

- Eco-construction;

1 https://youmatter.world/fr/innovation-technologique-resoudre-crise-ecologique/

- Alternative energy;
- Sustainable agriculture.

In each of these sectors, innovations considered environmentally friendly are examined through their characteristics and geographic areas of implementation. This analysis aims to contribute to ongoing debates surrounding sustainable development practices and their actual effectiveness in developing countries.

Justification of the Title

The choice of this research topic stems from a desire to address a cross-cutting issue that affects both rural and urban environments, and one that has the potential to enrich knowledge in key sectors such as housing, energy sources, and agriculture. Ecological innovations—often presented as both cost-effective and efficient—appear particularly relevant in the context of Senegal, where poverty levels remain high. Moreover, although a few dissemination projects exist, these environmentally friendly practices still seem relatively unpopular. As such, this study focuses more specifically on the challenges that hinder their widespread adoption.

It is worth recalling that several negative environmental changes have been observed in both rural and urban settings. Since 2005, participation in rain-fed agriculture in Sassar, a locality in the commune of Fissel Mbadane, Mbour department (Thiès region, Senegal), has made it clear that land degradation is visibly impacting agricultural yields. The traditional practice of spreading manure from domestic animals—an ancestral method—often helps improve harvests. However, due to the large number of cultivated plots, the available manure is insufficient to meet overall demand. Some families supplement this with chemical fertilizers, while others are unable to afford any due to economic hardship. In this context, it has been observed that chemical fertilizers alone do not offer a long-term sustainable solution. Without the addition of organic matter, soil productivity continues to decline year after year. Discovering that sustainable agriculture may provide a viable response—according to its advocates—sparked the idea of exploring this solution objectively through research.

The same reasoning was applied to the sectors of eco-construction and alternative energy.

Having access to adequate housing is a shared concern across all social strata, in both rural and urban areas. Traditionally, homes were built with locally available materials such as raw earth, straw, wood, or crop residues. With the widespread use of reinforced concrete (cement, sand, gravel, steel), a new face of housing emerged—particularly visible in urban development. While this shift has brought improvements in comfort and building durability or height, the fundamental issue remains cost, especially for youth and middle-income groups. For the poorest populations, access to such housing is often completely out of reach. In rural areas, disparities have emerged in the form of heterogeneous building types that reflect a tension between tradition and modernity—often corresponding to the household's economic status. Meanwhile, many existing permanent structures are roofed with metal sheeting (zinc), which performs poorly in extreme temperatures, especially during heatwaves. Similar challenges are visible in urban areas as well, though often to a lesser extent and with specific geographic variations (e.g., city centers, suburbs, peri-urban zones). When combined with other issues such as sanitation, it becomes clear that access to quality housing remains a pressing concern. According to its proponents, eco-construction offers appropriate responses to these challenges. But what exactly is eco-construction? What kinds of solutions does it propose?

In terms of energy, Senegal—as a sub-Saharan country—is facing an energy deficit. This situation has negative repercussions for the economy and poses a serious threat to the sustainability of already fragile ecosystems. In the face of limited access to modern energy, wood remains the primary source of energy for cooking. "It is estimated that it accounts for about 87% of household energy consumption in rural areas" (Le Partenariat, 2018). Another illustration of the pressure on ecosystems is the advancing desert in the north of the country, as well as the increasing rarity of certain wildlife species—such as the national emblem, the lion (the last lion on the Cap-Vert Peninsula is believed to have been shot by a hunter in...1905). Innovation in the energy sector is therefore also a clear and urgent priority.

It is from these observations and motivations that this research project was initiated. Personal experimentation with eco-construction methods was also undertaken in Fissel Mbadane (Commune of Mbour, Thiès region) in order to engage more directly with the practical aspects of the subject. One such initiative involved the challenge of transforming a traditional palisade into a permanent boundary wall. The following pages will present this process in greater detail.

Problem Statement

> "More than a green economy, Senegal envisions the emergence of genuine green governance—governance built upon balanced economic and social policies, supported by environmentally sound technologies and production methods. Reason compels us to do so."
>
> —Excerpt from the speech of His Excellency Macky Sall, President of the Republic of Senegal, delivered at the United Nations Conference on Sustainable Development, Rio+20 (Republic of Senegal, 2019)

This study is situated within two interrelated contexts: the broader global context and the specific national context of Senegal.

At the global level, current debates focus on critical issues such as housing, the energy crisis, renewable energy, and conventional agriculture. These are pressing topics on the international agenda. In the housing sector, new, more environmentally responsible construction methods are emerging worldwide. In the energy sector, renewable sources are increasingly being presented as viable alternatives to conventional fossil-based systems. In agriculture, significant shifts towards sustainability are being promoted through various certification schemes and labels.

At the national level, Senegal has seen a growing number of projects and programs aimed at improving the living conditions of its population. Over recent years, the promotion of ecological actions in partnership with international organizations has become a key component of national efforts to combat climate change, resource degradation, and poverty. For instance, in 2015, the National Domestic Biogas Program of Senegal (PNB-SN) was launched with the objective of constructing 10,000 biodigesters across the country during the 2015–2019 period. The program received financial support from both the Government of Senegal (1.679 billion CFA francs) and the European Union (4.85 billion CFA francs). Its main aim was to encourage rural and peri-urban households to adopt domestic biogas as a cleaner and more sustainable cooking fuel alternative to wood and charcoal—both cited as major contributors to deforestation in the country. This program was aligned with Senegal's Energy Sector Development Policy Letter (LPDSE) dated July 26, 2012, and the broader Plan for an Emerging Senegal (PSE). The overarching goal was to sustainably resolve the energy crisis by envisioning a "Senegal where electricity is available in sufficient quantity and quality, competitively priced, and produced using a diverse mix of technologies—including coal, gas, hydropower, wind, and solar energy." As part of this vision, the country aimed to reach a 15% independence rate in commercial energy (excluding biomass) by 2025, primarily through contributions from renewables and biofuels.

However, the PNB-SN had previously undergone a pilot phase from 2009 to 2013, which ultimately ended in failure. Of the 8,000 biodigesters initially planned, only 600 were constructed, and 30% of those were non-functional. Following this setback, the primary donor, the Dutch cooperation, withdrew its support, though the Senegalese government opted to continue the project. Unfortunately, the subsequent phase of the PNB-SN also faced serious implementation challenges, with many installations becoming non-operational. For example, in Notto Gouye Diama, a commune in the Thiès region, data collected in 2018 indicated that fewer than 30 biodigesters had been installed out of a target of 2,000 for the area. A similar situation was observed in Saint-Louis, where the NGO Le Partenariat was involved. Their joint objective with PNB-SN was to construct 2,000 digesters across the northern regions (Saint-Louis to Matam) through technical, institutional, and financial collaboration. Yet, as of 2018, only 65 digesters had been built in the entire zone—many of which were also non-functional.

Given these disappointing outcomes, the core issue identified is that in Senegal—as in many other so-called "developing" countries—ecological innovations often fail to achieve their intended objectives, despite promising theoretical frameworks and robust promotional strategies.

From this overarching problem arises the fundamental research question: **why do ecological innovations in countries like Senegal so frequently experience failure in practice, despite their conceptual promise and international support?**

This central question generates several sub-questions, including:

- How can we characterize the diffusion process of ecological innovations, particularly in terms of their limited spatial spread and lack of long-term adoption?

- To what extent do the technical, financial, social, or other aspects of ecological innovations align with the socio-territorial realities of the target areas?

Hypotheses

Two main hypotheses underpin this research:

- Hypothesis 1 (H1): Certain ecological innovations may be accessible and capable of improving the living conditions of populations in various sectors; however, they suffer from limited diffusion. This results in low spatial adoption and a lack of sustainability in the few areas where they have been implemented.

- Hypothesis 2 (H2): Other ecological innovations, by contrast, may contain technical, financial, social, or other elements that are incompatible with the contextual realities of host territories, preventing their widespread adoption.

These hypotheses give rise to the two primary objectives of this study, which are:

- To verify the validity of these assumptions through a comprehensive presentation of key sectors and selected case studies in Senegal.

- To better understand the dynamics behind the implementation and diffusion of ecological innovations in the country.

Significance of the Study

What value does this topic bring to geographical thought, scientific inquiry, and society at large?

"A problem identified is already half-solved."

This research is grounded in that guiding principle. Its purpose is to identify the major challenges hindering the implementation and/or diffusion of ecological innovations in Senegal. Furthermore, it seeks to go beyond the narrative frameworks commonly used by NGOs, development partners, and public authorities. The study aims to contribute meaningfully to ongoing debates: Are we dealing with illusory promises for "underdeveloped" countries, or are these truly viable pathways toward tangible social, economic, and environmental gains?

From a geographical perspective, the study aligns with the long-standing tradition of examining the relationship between humans and their environments—an enduring concern of the discipline. As early as 1923, Barrows defined geography as "human ecology." Environmental issues have always played a role in geographical inquiry, initially through deterministic approaches that emphasized environmental influence on human activities, then through more reciprocal analyses leading to the environmentalist paradigm, and eventually through the study of eco-development at the global scale (Belhedi, 2010). In the 21st century, the magnitude of challenges related to sustainable development, climate change, and renewable energy makes it impossible for geographers—particularly those in the Global South—to remain indifferent.

In this context, a focus on the challenges of ecological innovations in Senegal, through the lenses of eco-construction, alternative energy, and sustainable agriculture, is both timely and justified.

From a scientific perspective, this study seeks to address the lack of independent research on ecological practices in Senegal. Most existing publications in this field are commissioned by public authorities, NGOs, or development partners. As such, this work can serve as a complementary bibliographic resource, enriching the knowledge base related to the implementation and monitoring of ecological projects in the region.

From a societal standpoint, this book may offer insights for improving both environmental conditions and the daily lives of local communities, by identifying the principal obstacles that undermine the success of ecological initiatives. A clear understanding of these barriers could inform the development of adaptive strategies to achieve better outcomes.

Methodology

> "If I want to understand a machine, I will disassemble it to study each part individually. Once I have an accurate understanding of each component and can reassemble them in their original order, I will then perfectly comprehend the machine, for I will have deconstructed and reconstructed it." — Condillac

The methodological framework outlines the research process and experience. It describes the necessary steps to verify the hypotheses derived from the theoretical framework (Gauthier, 1997). As Lussier (1997) emphasizes, it includes all the information that would allow others to replicate the study, retrace the steps, and verify the conclusions (Dubois, 2004). This framework encompasses the different phases of investigation, the tools for data collection and analysis, and also the constraints and challenges encountered throughout the research process.

In this study, the methodology is structured into three subsections:

- Literature and documentary research;
- Definition and delimitation of study areas (field sites);
- Collection, analysis, and processing of field data.

• Documentary Research

"If you steal from one author, it's plagiarism; if you steal from many, it's research."
— Wilson Mizner

The research began with documentary investigations of varying intensity. Given the large number of ecological innovations in the targeted sectors worldwide—and the constant emergence of new technologies ranging from low-tech to high-tech solutions—the main challenge was identifying those already present in Senegal and suitable for representative analysis. This identification process required time, involving both desk research and exploratory fieldwork.

Initial literature reviews focused on the global scale to provide an overview of the latest ecological innovations. For instance, in the field of alternative energy, it was discovered that plants can generate electricity. This early-stage innovation was patented in 2007 in the Netherlands by researchers at Wageningen University. The commercial enterprise, Plant-e, captures electrons via electrodes placed near plant roots. Since plants produce more energy than they consume, the system's advantage—compared to wind or solar—is that it can function at night and during windless periods (Futuraplanete, 2015). Similarly, in 2016, three Andalusian students in Spain founded the startup Bioo, which develops a biological battery that uses photosynthesis to generate electricity.

Such examples demonstrate the rapid evolution of ecological innovations, reinforcing the need to examine the Senegalese context more closely—where many of these technologies are either not well developed or not yet implemented. To this end, several emails were sent to key organizations such as: La Voûte Nubienne (contact@lavoutenubienne.org), focused on earthen construction; Innovation Environnement Développement Afrique – IED Afrique (contact@iedafrique.org), which promotes local agricultural innovations; Malem-Auder ASBL (asbl@malem-auder.org), a Belgian-Senegalese association working in sustainable construction and agroecology.

The first field visits took place in August and September 2019, at the headquarters of Enda Tiers-Monde in downtown Dakar and IED Afrique in Liberté 6, Dakar.

The innovations ultimately selected for in-depth study in each sector were:

- Eco-construction: Nubian Vault and Compressed Earth Blocks (CEB or BTC);
- Alternative energy: Biogas and biochar (with biogas later replaced by the Taiba Ndiaye wind farm, already under study);
- Sustainable agriculture: Biofertilizers and urban micro-gardening.

A wide range of sources were consulted, including general and specialized books, academic theses and dissertations, scientific journal articles, program fact sheets, as well as multimedia sources such as videos, forums, and audio content. Due to the volume of materials (exceeding 10 pages), a summary presentation of key works was chosen over an exhaustive listing, to avoid overloading the general introduction.

• Delimitation of Study Areas (Field Sites)

Once the innovations were selected, corresponding geographical areas were identified for field study. A sector-based geographical approach was adopted. The goal was not to study all of Senegal—an overly broad undertaking—nor to focus on a single zone or sector, but rather to analyze targeted actions within defined areas. For each innovation, a key diffusion actor was chosen, and field visits were conducted in their operational locations. For some technologies with broader implementation—such as biofertilizers—a mix of documentary research, telephone interviews, and agricultural sector analysis was conducted in lieu of fieldwork.

The selected zones, based on the diffusion actors, were:

Podor (northern Senegal) – Nubian Vault

Taiba Ndiaye (central-western Senegal) – Wind Farm

Toubacouta (near The Gambia) – Biochar

Dakar – Urban micro-gardening and part of the CEB case

Mbour (Gandigal) – Second CEB case study

Due to time constraints, research on biofertilizers was limited to literature review, telephone interviews, and a comparison with the realities of agriculture and rural life in Senegal.

The following figure provides a visual summary of the initial methodological elements:

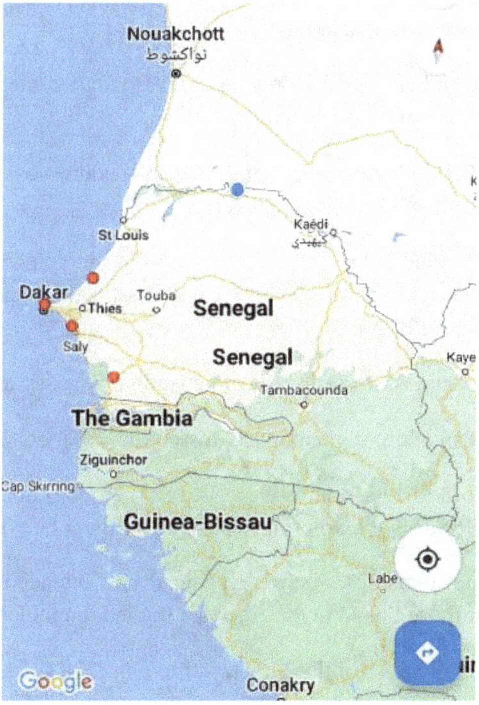

Fig. 1. Research areas
Source: Author, 2022

The blue dot represents the GPS location at the time of the screenshot on June 14, 2022, in Podor.The other red dots were added later; they represent other areas involved in the study (several field visits were conducted in Dakar). From top to bottom, the zones are: Podor (Nubian vault), Taiba Ndiaye (wind farm), Dakar (urban micro-gardening and CEB), Mbour (CEB), Toubacouta (biochar).

Field Data Collection, Analysis, and Processing

The collection of field data was carried out on different dates and through several methods. In addition to direct observation, interview guides, questionnaires, a focus group, and several informal discussions were conducted or developed. The goal was to gather information about the innovations, their characteristics, the beneficiaries, and the dissemination actors in order to understand the challenges and verify the research hypotheses.Ten interview guides were developed, including:

- 1 on the Nubian vault, addressed to LePartenariat, the NGO managing this program in northern Senegal;
- 2 on CEB (Compressed Earth Blocks), addressed to Elementerre and Village Pilote, key actors in this field;
- 2 on the wind farm: one addressed to the municipality of Taiba Ndiaye, where the innovation is located, and one to the company managing the wind farm (PETN);
- 1 on biochar, addressed to the organization Nebeday in Toubacouta;
- 1 on urban micro-gardening, addressed to the City of Dakar;
- 3 on biofertilizers, addressed to Biotech Services Senegal, Éléphant Vert, and AVF-Senegal (exploring the use of fertilizer trees).

As for the questionnaires, six were planned and intended for the beneficiary populations or those linked to the practices studied, including:

- 1 for beneficiaries of the Nubian vault in the Fouta region;
- 1 for beneficiaries of CEB buildings or infrastructure;

- 1 for biochar beneficiaries in Toubacouta;
- 1 for residents of Taiba Ndiaye affected by the wind farm;
- 1 for biofertilizer users;
- 1 for urban micro-gardening beneficiaries in Dakar.

A focus group was conducted on May 4, 2022, with the women's association of CFD/CPC Patte d'Oie in Dakar, concerning micro-gardening in a space set up by the City Council. Table 1 shows the participants.

Table 1: Focus Group Participants on Urban Micro-Gardening in Dakar

Participant	Role
Saly Thiao	President of the Women's Group at CFD/CPC Patte d'Oie
Ndeye Coumba Sall	Beneficiary at CFD/CPC Patte d'Oie
Khady Fall	Beneficiary at CFD/CPC Patte d'Oie
Awa Ndao	Beneficiary at CFD/CPC Patte d'Oie

Source: Author, 2022

An informal interview was also conducted on the same day, May 4, 2022, with the main manager of the CFD/CPC of Grand Dakar, in Dakar. In addition to field visits carried out in 2019, further investigations took place on the following dates: 03/04/2022, 30/04/2022, 03/05/2022, 14/06/2022, 07/07/2022, and 16/11/2022. During these periods, several sites were visited, including:

- the CEB production site of Village Pilote in Deni Biram Ndao, on the outskirts of Dakar;
- the wind farm in Taiba Ndiaye, located in central Senegal (Thiès region);
- the Nubian vault actors in the north of the country, in Podor;
- the biochar actor (Nebeday) near the Gambian border in Toubacouta (Fatick region);
- and the CEB production site of Elementerre in Gandigal (Mbour).

During the field trips, smartphones were extensively used (for phone calls, photography, research, geolocation of study areas, etc.). As most of the terrain was unfamiliar, GPS was especially useful as a navigation tool during the fieldwork, as illustrated in Figures 2 and 3.

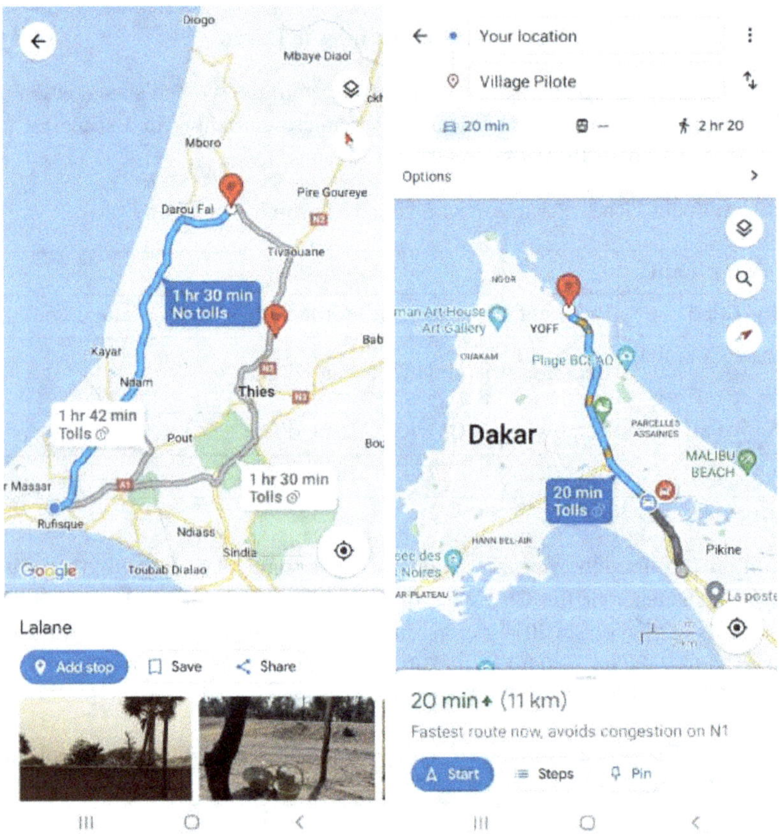

Fig. 2 & 3: Travel Techniques to Reach the Study Areas
Source: Author, 2022

Figure 2 shows the route taken on the first trip to Taiba Ndiaye from Rufisque on May 3, 2022, while Figure 3 shows the route to the headquarters of Village Pilote in Yoff on May 4, 2022. In addition, participation in the census mapping work for Senegal's Fifth General Population and Housing Census (RGPH-5) organized by ANDS (from March 2022 to January 2023) made it possible to further explore the study sectors in connection with the geography of Senegal, through a series of direct observations. These included:

- A tree nursery project in Ndiob, a commune in the Fatick region;
- Traditional earthen constructions;
- A mini solar power plant in Bacco Dior (Fatick);
- Agricultural practices during both dry and rainy seasons;
- And general living conditions of parts of the population in various geographic zones.

Data Processing and Analysis

Data analysis initially took the form of a comparison—between how the innovations were presented, their intended objectives or potential impacts, and the actual realities observed in the field. Therefore, the qualitative approach was predominant, although elements of quantitative analysis were also incorporated.

Conducted Experiment Eco-construction Experiments in Fissel Mbadane

Eco-construction trials were carried out in Fissel Mbadane. The first experiment aimed to meet the challenge of turning a traditional palisade into a perimeter wall. Already in 2018, an attempt had been made by applying a raw earth plaster mixed with millet husk to a palisade made of cemented poles. Unfortunately, due to the wide spacing of the poles and heavy rains, the structure collapsed. In March 2019, the idea was revisited—this time replacing the earthen plaster with cement and placing the eucalyptus wood poles closer together, reinforced with concrete. That wall withstood the 2019 rainy season, but not the one in 2020. The wood used was not sufficiently resistant to water load and was weakened by termite attacks. Finally, in 2021, a thicker wall was built using a mixture of raw earth mortar and rubble, with some solid bricks added to a portion. The top of the wall was cemented to prevent water-related issues. For another section, a mason was hired to build a conventional wall with reinforced concrete posts and tie beams.

Why These Experiments?

These experiments were carried out to find alternatives to traditional palisades, which lack durability over time, and also to solid fences, which are often too expensive for many residents. The aim was to find a compromise between the two: a weather-resistant fence at a very affordable cost. This led to the idea of starting with raw earth plaster, a free and eco-friendly material. The project also provided an opportunity to apply academic knowledge in practice. However, the increased weight during the rainy season, combined with strong winds, had not been fully anticipated.

Lessons Learned

These experiments led to several valuable insights:

First, ideas that once seemed purely theoretical were put into practice.The first trial showed that earth adhered well to millet husk and became hard after drying. The main challenge was keeping the palisade standing in the face of wind and rain. After several failures, it became clear that very strong, weather- and termite-resistant poles were needed (since termites also attack millet husk). A sloped cap (to direct rainwater) would likely be very effective when used with raw earth plaster.

Photo 1: Example of a traditional palisade made with a type of reed called "mbourndou" in Serer.

Source: Author, 2019, Sassar / Fissel Mbadane / Mbour / Senegal

This type of fence can last 2 to 3 years, but it requires more time to build and a large number of wooden posts.

Photo 2: Another type of traditional palisade made with millet husks.

Source: Author, 2018, Sassar / Fissel Mbadane / Mbour / Senegal

This palisade is called "kéb" in Serer and is the most common form. Its implementation is quick, but its lifespan is short (about 1 year if exposed to rain).

Photos 3 and 4: Glimpses of the first experiment using raw earth plaster.

Source: Author, 2018, Sassar / Fissel Mbadane / Mbour / Senegal

The tilting caused by the weight can already be seen in the background. The corner wooden post was not very sturdy.

Photos 5, 6, 7: Glimpses of the second experiment using cement plaster.

Source: Author, 2019, Sassar / Fissel Mbadane / Mbour / Senegal

Cement also adheres well to millet husks. For this experiment, about 10 bags of, cement, nails, wire, and eucalyptus wood were purchased.

Photos 8, 9, 10: Glimpses of the third experiment and the conventional wall.

Debris and rubble found on-site were gathered to build a wide wall joined with raw earth. It is very sturdy and withstood the 2021 rainy season without any issues.

Source: Author, June 2021 (Photo 8), January 2022 (Photos 9 and 10), Sassar / Fissel Mbadane / Mbour / Senegal

Conceptual Framework and Analysis

This section provides an explanation of the various key concepts related to the topic, sometimes along with the theoretical perspectives surrounding them.

Innovation, Ecological Innovations, Ecology, Sustainable Development

Innovation refers to the idea of novelty "in a given time and place" (Brunet et al., 1993). Its etymology comes from the Latin in (into) and novare (to make new, renew, restore, transform, change). It is distinct from invention or discovery due to its practical nature and concrete implementation (www.toupie.org). Fournier (2019) considers innovation as an improvement of something already existing. Similarly, the digital dictionary 38 Dictionnaires et Recueils de Correspondance links it to the transitive verb "*to innovate,*" meaning: "*to introduce something new into a technique, an organization, etc.*" According to Cécile et al. (2001), innovation is an economic process involving variables such as profitability, competitiveness, investment, etc. They also define it as a social process (involving actors, converging or diverging interests, individual and collective representations), and as an institutional process that occurs within programs and organizations—public or private—and involves negotiations and decision-making procedures.

Two types of innovation can be distinguished:

- Incremental innovations: These are improvements made to products and/or production techniques, aimed at enhancing quality, productivity, or diversity. They occur continuously throughout the history of technology. While they do not create major disruptions, they are essential to explaining productivity gains, market expansion, and trend dynamics (Op. Cit.).
- Radical innovations, on the other hand, represent breakthroughs in the evolution of processes or products, leading to major transformations in production methods, marketing strategies, and required professional skills. These do not occur continuously over time. The convergence of multiple correlated radical innovations can lead to the emergence of a new technical system (Ibid.).

By definition, innovation is considered positive, at least in one specific aspect of a problem or situation. However, that does not mean all innovations are entirely beneficial. A critical look at technological innovation over the past centuries reveals that although they have generally had very positive outcomes, particularly in computing, medicine, energy, and transportation—improving our lives, health, productivity, and comfort—many of them are also at the root of our ecological problems. For example: Energy technologies are among the main sources of greenhouse gases. Chemical innovations have greatly contributed to polluting ecosystems and affecting biodiversity. Innovation and mechanization in agriculture are also responsible for the degradation of soil quality. In fact, the vast majority of current environmental issues are the direct consequences of technological innovations over the last 150 years (Fournier, 2019).

In some cases, the term ecological is added to "innovation" to describe its positive environmental impact. Ecological innovations are considered approaches that promote environmental protection, aligned with the goals of ecology. As the science of ecosystems, ecology studies the *"relationships between living beings and their environment."* Over the years, "*it has come to encompass all concerns related to managing humanity's habitat, returning to its Greek etymological roots. Thus, ecology is simultaneously: A milieu (environment), A science, and A set of practices or even norms*"(Brunet et al., 1993). Defined this way, ecology covers a very broad scientific field, drawing on a variety of disciplines such as: Climatology, Hydrology, Oceanography, Chemistry, Geology, Pedology, Mathematics, Physiology, Genetics, Ethology, (Microsoft Encarta, 2008).

Ecological innovations do not always require heavy investment. They may result from clever combinations of existing techniques. These innovations align with the framework of sustainable development, which promotes a model of economic, social, and environmental growth that the planet can support in the long term. Today, the notion of sustainable development is highly popular, but its origins go back several decades. The evolution of the concept of development has been marked by several key dates:

- Late 1960s: At this time, the first signs of awareness began to emerge regarding environmental problems caused by the pursuit of economic growth. Rapid industrialization and the exploitation of natural resources started to reveal their limitations, highlighting negative environmental impacts such as ecosystem degradation and pollution.

- 1972: The United Nations Conference on the Human Environment marked an important turning point by emphasizing the scarcity of natural resources. This conference urged states to take measures to limit the consumption of these resources, thus laying the groundwork for discussions on the need for more environmentally respectful development.

- 1987: The Brundtland Report, written by the Brundtland Commission, represents a decisive milestone in the history of sustainable development. This report painted an alarming picture of the planet, highlighting major threats associated with the logic of unilateral economic development. Among the main concerns were: the extinction of many plant and animal species; the population explosion in developing countries; air and water pollution; major ecological disasters such as the Chernobyl nuclear accident in 1986; and the worsening social conditions of populations in the Global South. In response, the report called for the pursuit of "another kind of development" that meets people's needs without destroying the natural resources on which those needs depend. It is within this context that the concept of sustainable development emerged. It is based on the principle of meeting the needs of the present without compromising the ability of future generations to meet their own. The idea is to balance economic, social, and environmental priorities to ensure a viable future.

- 1992: The Earth Summit in Rio de Janeiro was a key moment in the international recognition of sustainable development. At this summit, 182 states signed commitments to sustainable development and adopted the Rio Convention on the conservation of biological diversity. The summit also saw the birth of Agenda 21, an action plan for the 21st century that outlines concrete measures for promoting sustainable development on a global scale.
- 1998: The Kyoto Protocol paved the way for formal international commitments to combat climate change. Coming into force in 2005, this protocol required the reduction of greenhouse gas emissions—the main drivers of global warming. It marked a major turning point in the fight against climate change.
- 2002: The World Summit on Sustainable Development in Johannesburg, held ten years after Rio, served to assess progress and identify ongoing challenges. This summit emphasized the need to intensify efforts to achieve sustainable development goals, particularly in the areas of poverty, water, energy, health, and biodiversity.

However, on the other hand, the concept of sustainable development can be problematic, especially in underdeveloped countries. This complexity largely stems from the need for these countries to reconcile and arbitrate between objectives that operate on different timeframes: economic growth as a short-term driver of progress, and sustainable development as a means to support livelihoods in the medium and long term. Additionally, the theoretical separation of the economic, environmental, and social sectors in the sustainable development framework often results in a practical focus on one sector to the detriment of the others (Kaere, 2012).

In addressing the sustainability of development, ecological innovations are considered suitable solutions. However, to be effective, an innovative approach must combine with and adapt to the social, commercial, and technical realities of the context in which it is implemented (Treillon, 1992). Therefore, it must overcome conceptual, organizational, knowledge-based, technological, and especially financial barriers. Consequently, innovation in itself is not a source of change—its diffusion process is. This process has two fundamental components:

- The involvement of change agents, and
- The perceived characteristics of the innovation by the adopters.

The role of the change agent is multifaceted in relation to potential adopters: making contact, persuading, supporting, identifying needs, and showing empathy. Regarding the features that the innovation must present to potential adopters, Rogers highlights the importance of factors such as: relative advantage, compatibility, complexity level, the use of demonstrations, and the possibility of trials (Cécille et al., 2001).

Eco-construction or Green building

> "Eco-construction, bioclimatic design, ecological housing... several names for the same philosophy." (Peyrot and Istin)

> "Eco-construction, or sustainable building, refers to the creation, restoration, renovation, or rehabilitation of a building while ensuring it respects ecological principles as much as possible at every stage of its construction—and later, in its use (heating, energy consumption, waste and water management)." (Wikipedia, 2020)

Eco-construction (or green building) is an approach aimed at limiting the environmental impacts of construction while improving the economic and health aspects for its users (Plateform-ecco2). According to Peyrot and Istin, a house is considered well-designed when it is well-oriented, well-integrated into its environment, adapted to its climate, low in energy consumption—or even energy-producing—and built using locally sourced materials chosen based on their physical properties.

This philosophy of eco-construction, which is said to exist in nature itself (e.g., termite mounds), dates back to ancient times in various civilizations around the world. The Nubians of the Nile were already building vaults out of raw earth more than 3,000 years ago. The Greeks used an underfloor heating system called the hypocaust as early as the 6th century BCE; the Persians had their wind towers, or badgirs. In the Middle Ages, similar housing concepts existed in various regions, such as in Europe with half-timbered houses (wooden frames filled with raw earth bricks and covered with plaster or wattle and daub) (Peyrot and Istin).

The modern eco-construction movement began in the wake of the 1970s oil crisis, which pushed for new construction approaches. According to Wikipedia (2020), its beginnings can be traced back to the late 1940s. Peyrot and Istin, however, place its origins in the 1960s in the United States, with an architectural movement inspired by the nature-integrated works of architect Frank Lloyd Wright (early 20th century). This movement stood in radical opposition to conventional construction, instead promoting the use of natural, local, and traditional resources, as well as renewable energies such as sunlight. These early advocates of alternative construction were generally self-taught and had limited means, but embraced a degree of detachment from political power. Experiments multiplied, gaining momentum after the oil crisis and spreading to Europe in the early 1980s. These years of experimentation primarily focused on individual or collective housing, testing both non-standardized materials and traditional methods such as rammed earth, straw, adobe bricks, etc.

Since then, several principles of bioclimatic design have been refined, including:

- Local climate conditions and environmental adaptation. In a cold region like Siberia, the more compact a building is, the easier it is to heat. In contrast, in an arid region, it is important to promote cool air flow inside the house, using materials with high thermal inertia, such as raw earth, for example (Peyrot and Istin).

- The shape and orientation of the building, since walls serve one or several functions depending on their geographic positioning. A wall can capture heat, lose heat, or even do both, depending on its orientation and the role of the roof (Op. Cit.). Therefore, one must design with the site's topography and climate in mind, considering two essential parameters: sun exposure and compactness.

- The functions of windows or glass doors are multiple: they allow light to enter, provide a view of the outside, and serve for ventilation. From a thermal perspective, they allow heat to be captured depending on their location/orientation. From this point of view, they are more recommended in cold and temperate zones (Peyrot and Istin).

- Ecological materials and equipment. These are ecological construction and insulation materials such as stone, raw earth bricks, hemp, straw (straw bale houses), wood, wood fiber, sheep's wool, cellulose wadding… (Wikipedia, 2020). The use of materials whose raw materials require high energy for extraction, manufacturing, or transport is, on the other hand, not recommended (Op. cit.). For timber frame construction, it is preferable that the wood comes from a certified source. To further improve insulation, it is also possible to use a green roof or a vegetated wall. In addition to the energy savings gained through good insulation, the use of renewable and natural energies such as passive solar energy, photovoltaic panels, solar water heaters, wind turbines, hydropower (hydraulic turbines, water mills), and all energy sources derived from biomass—such as wood, biogas, composting—can also help reduce the ecological footprint (Wikipedia, 2020).

- The management of rainwater and wastewater is also part of the principles of bioclimatic construction. Collecting rainwater in buildings is increasingly seen as essential, as it helps save drinking water from the main network. Its use is particularly suitable for purposes such as toilets, laundry, or watering green spaces, according to Peyrot and Istin. Ecological sanitation is part of this recycling approach while respecting the natural cycle of organic matter. It is possible to separate and treat blackwater (from toilets) independently by composting it to produce enriched organic inputs for the soil, and greywater (from household residues) by setting up a phyto-purification system. This type of sanitation relies on the purifying power of plants and their root micro-ecosystem, combined with mechanical filtration.

- Pretreatment is carried out in a vertical-flow planted filter, composed of several purifying plants placed on a mineral substrate (pozzolan, sand, etc.), which retains solid matter on the surface while allowing water to flow through. The organic matter decomposes without emitting bad odors or producing sludge. The treatment continues in a second horizontal-flow planted filter, where dissolved organic matter is gradually broken down. The plants absorb nitrogen and phosphorus contained in the wastewater before it is directed to an outlet, such as a pond or a willow plantation (Peyrot and Istin).

- Maintenance: The notion of upkeep and maintenance is closely linked to eco-construction. Its implementation is primarily intended to protect biodegradable materials in order to extend their lifespan. Construction defects, as well as wear and tear caused by occupants, are factors that can lead to rapid degradation (ILO, 2013). In modern construction, risks associated with poor maintenance are lower thanks to the properties of concrete. However, they are not nonexistent, and when repairs are necessary, they are often more complex and costly. Therefore, in an "eco-designed" house, occupants must be particularly attentive, requiring regular and cyclical maintenance (repointing and repair of protective coatings, painting or wood treatment, termite protection, covering straw during the rainy season, etc.) (Op. cit.). Eco-construction cannot be sustainable without constant maintenance, which ensures the preservation of the mechanical, thermal, acoustic, lighting, and aesthetic performance of materials. Moreover, educating occupants on proper cleaning practices—especially regarding water use—helps prevent inconsistencies that could affect the building's durability (ILO, 2013).

- Dismantling: The relative fragility of natural materials throughout their use as construction products is a major advantage at the end of their life (ILO, 2013). Their dismantling mainly generates biodegradable waste, which is easily absorbed by nature (Op. cit.). In contrast, concrete construction, besides producing rubble, also generates many other inert wastes: tiles, aluminum, iron, paint, etc. While recycling solutions exist for these materials, the processing capacity remains limited—except for metals and glass (ILO, 2013).

Alternative Energies

Alternative energies are those that differ from fossil or nuclear energy and are seen by their proponents as potential replacements. They are also called renewable energies, due to the inexhaustible nature of some (such as wind and sunlight) or their rapid renewal (like biomass). Given the challenges of conventional energy (scarcity, environmental issues, etc.), alternative energies have enjoyed widespread support and a strong reputation in the global energy landscape—partially explaining their worldwide growth. Already in 2011, they accounted, according to BFM BUSINESS (2013), *"for nearly half of the installed electricity capacity worldwide"*. Examples include: wind energy, solar energy, hydroelectricity, geothermal energy, and biomass energy.

- Wind energy: This comes from the movement of air masses across the globe. Humans have used wind for sailing since ancient times. Later, windmills exploited this resource. Today, modern wind turbines convert the wind's mechanical energy into electricity. A distinction is made between onshore wind farms built on land and offshore wind farms built off the coast (Ibid.).

- Solar energy: There are two main techniques for generating electricity using sunlight and heat: solar thermal energy and photovoltaic solar energy. In the case of a solar thermal power plant, solar radiation heats a fluid that is then transformed into steam. This steam powers turbines to produce electricity (green-economy.fr, 2020). In the case of photovoltaic solar panels, the cells react to sunlight by producing direct current electricity (Op. Cit.).

- Hydroelectric energy or hydropower: This is energy derived from moving water. It has been used since ancient times, for example with water mills (green-economy.fr, 2020). Today, it includes renewable electricity production (waves, tidal energy, hydroelectric dams). In the case of dams or pressurized pipes, the force of the water drives turbines to generate electricity (Ibid.).

- Geothermal energy: This taps into the heat produced by the Earth's core. The thermal energy from hot springs or underground reservoirs is used directly or converted into electricity (green-economy.fr, 2020). Water can also be injected deep into the ground, where it heats up upon contact with hot rocks and rises back loaded with thermal energy (Op. Cit.).

- Biomass energy: This is solar energy stored in organic form via photosynthesis (wood and biofuels). Biomass is considered one of the most promising geoengineering techniques in the IPCC's October 2018 report. This technique, called BECCS (Bioenergy with Carbon Capture and Storage), involves growing vegetation, burning it to produce energy, and capturing the CO_2 released during combustion to store it in geological reservoirs (like former gas or oil fields) (green-economy.fr, 2020).

However, despite these energy options, criticisms are often raised against renewable energy. The classic debate revolves around their low efficiency and the intermittency of some sources—meaning they function intermittently, such as wind turbines when wind speeds are low, or solar energy at night. According to Contrepoint (2019), the term "alternative energies" is frequently interpreted as referring to energies that are different from traditional sources like fossil fuels or nuclear power. However, "alternative" should be understood in the same sense as "alternating current": sometimes positive, sometimes negative. According to the author, "*placing wind turbines across the countryside simply adds unsightly windmills to the random production of traditional power plants rather than replacing them.*" The following figure is provided for illustration.

Fig. 4. Old and New Energy
Source: www.contrepoints.org

One can observe the difference in the landscape, with the disappearance of birds in the image on the right, in addition to the space occupied by the new wind turbines.

For www.green-economy.fr (2020):

> *"Renewable energies are actually far from being clean, and even less 'renewable'. Wind, sunlight, and the movement of water (river currents, waves, marine currents, etc.) are indeed inexhaustible—at least on the scale of our civilization. But that is far from being the case for the materials used to convert this energy into electricity."*

In the case of biomass, the critique centers around the risks of biodiversity loss (habitat destruction, soil erosion, pollution, scarcity of water resources) and the threat to food security.

Sustainable Agriculture

> "Sustainable agriculture" is one of those phrases that is often used, but whose meaning is not always very clear! There's no official set of standards, no certification (unlike organic farming), and no legal text to define its practices—so what does it really mean? It's more of a mindset and a holistic approach than a strictly regulated method of production (fairebien.com/agriculture-durable, 2020)

Sustainable agriculture is the application of the principles of sustainable development to farming. It is a system of agricultural production that aims to ensure the long-term production of food, wood, and fibers while respecting ecological, economic, and social limits that ensure this production is maintained over time (Wikipedia, 2019). It is considered a modern return to the principles of traditional agriculture, which preserved its resources, recycled its waste, and protected its seeds and species. According to www.agridurable.wordpress.com (2020), the goals of sustainable agriculture are:

- To meet the needs of the population;
- To ensure sufficient production, with a focus on quality rather than yield;
- To reduce wealth inequality;
- To maintain environmental balance and preserve biodiversity.
- Taking livestock, fishing, and crops into account, sustainable agriculture thus aims for "an improvement in the sustainability of the system by creating more long-term wealth per unit of production, on a more equitable basis" (Wikipedia, 2019).

However, although sustainability is based on the targeted use of natural processes, the concept does not prohibit the use of chemical fertilizers and pesticides. Still, the need to purchase such products—along with fossil fuels for machinery—reduces sustainability, as it leads to dependence on non-renewable resources.

We can therefore conclude that there are no specific "sustainable agriculture techniques," but rather a desire to respect the core principles of how a natural ecosystem functions. According to www.geo.fr (2020), sustainable agriculture should strive to follow these criteria as much as possible:

- Optimal use of natural resources, especially water;
- Recycling of plant and animal waste to fertilize and maintain soil quality (compost and manure);
- Use of green waste as biomass (fuel, biofuel, biogas) to create energy;
- Limiting greenhouse gas emissions, especially by promoting short supply chains;
- Reducing pollution by lowering the use of fertilizers and pesticides;
- Maintaining and utilizing natural predators and pollinators;
- Product traceability to ensure food safety;
- Preservation of biodiversity, natural ecosystems, and the genetic heritage of endemic crop species;
- Agricultural landscape planning and combating desertification;
- Respect for animal welfare;
- Respect for working conditions and health of workers and local residents;
- Local economic development.

In practice, several forms of agriculture are considered part of sustainable agriculture:

- Peasant agriculture: Distinguished by its promotion of family farming, this form is not officially regulated. Nearly half of the world's population—around 3.5 billion people—are said to be peasants. This model aims to allow farmers to earn a decent living from their work and defends small-scale farms against large industrial operations such as "mega-dairies."
- Organic agriculture: Covers perhaps the widest range of practices. To farm organically means not using synthetic inputs or petrochemical-based pesticides. However, many organic farmers, apart from this exception, operate in ways very similar to conventional farmers: large-scale field cropping, in rows, on bare soil, with a lot of tilling, low biodiversity, and heavy mechanization.
- Agroecology: Goes beyond organic farming. In addition to techniques like composting, intercropping, or mound cultivation, it seeks to incorporate

ecological management principles into farming, including water efficiency, erosion control, hedgerow planting, and reforestation. Agroecology is both an agricultural technique and a scientific discipline and grassroots movement in defense of agricultural workers. Agronomy lies at the heart of agroecological systems. Strong knowledge in this field is deemed essential for both farmers and advisors. Every transition to an agroecological production system must be tailored to the specific territory (soil and climate conditions, socio-economic context) and the farmer's goals (such as quality of life)[2].

- Permaculture: Not strictly an agricultural system, permaculture is about designing sustainable, resilient human environments. It incorporates best practices from organic farming and agroecology, but also renewable energy, eco-construction, etc. Its applications are broad (cities, especially "transition towns"; businesses; energy systems; etc.). The spirit of permaculture is to link all elements of a system—including people—together. It seeks to replicate the diversity and interdependence of natural ecosystems to ensure health, efficiency, and resilience for every component and the system as a whole. It is a closed-loop system where each element supports others, producing no external waste. In its agricultural application, permaculture often draws inspiration from forests, where the soil is left undisturbed.
- Agroforestry: With trees at its core, agroforestry combines trees with crops or livestock on the same land. Trees draw water and minerals from deep soil layers and make them available to surface crops. They also protect crops during droughts and provide shade and shelter from heat. Agroforestry is seen as a solution to mitigate climate-related challenges that have contributed to stagnant cereal crop yields. Additionally, trees support organic activity and restore soil fertility through root and leaf decomposition[3].
- Integrated agriculture: While it doesn't entirely eliminate synthetic pesticides, it aims to significantly reduce their use. Integrated production uses pesticides only as a last resort, after implementing other methods such as biological control (e.g., using ladybugs instead of insecticides to combat aphids).
- Precision agriculture: A set of techniques and practices aimed at minimizing input use by adapting them precisely to plot variability—

2https://www.colibris-lemouvement.org/magazine/permaculture-agroecologie-agriculture-bio-quelles-differences
3https://www.colibris-lemouvement.org/magazine/permaculture-agroecologie-agriculture-bio-quelles-differences

adjusting doses based on quantity, timing, and location (modulation). It uses data collected by on-field equipment (e.g., GPS, weather stations) or remote sensing (e.g., drones, satellites), processed by onboard software to control equipment (spreaders, sprayers, seeders, etc.). Data are stored and analyzed to monitor farm operations. This approach is also used in precision livestock farming (e.g., automated concentrate feeders)[4].

After this general conceptual presentation of key study terms, it is now time to delve more specifically into the case studies in Senegal.

4 https://dicoagroecologie.fr/encyclopedie/agriculture-de-precision

Part One

Green Building, Alternative Energies, and Sustainable Agriculture in Senegal:

Characteristics and Geographical Challenges

Introduction to Part One

This first part addresses green building, alternative energies, and sustainable agriculture in Senegal, focusing on their characteristics and geographical challenges. It is divided into three (3) chapters: the first covers green building, the second alternative energies, and the third sustainable agriculture (in Senegal). Before discussing these chapters in detail, it is necessary to provide a brief physical and human overview of Senegal.

Senegal is located at the westernmost point of Africa, between latitudes 12°8 and 16°41 North and longitudes 11°21 and 17°32 West. It covers an area of 196,722 km² and is bordered to the north by Mauritania, to the east by Mali, to the southeast by Guinea, to the south-southwest by Guinea-Bissau, and to the west by the Atlantic Ocean. The country is divided into 14 regions, 46 departments, and 557 communes. It also includes an enclave, The Gambia, which extends more than 300 km inland.

• Climate and Vegetation

Senegal has a tropical Sudanian-Sahelian climate with two main seasons: a dry season (from November to May), marked by the predominance of maritime trade winds (in the northwest) and continental winds (inland). And ahot and rainy season (from June to October), dominated by monsoon winds originating from the St. Helena Anticyclone. Average annual rainfall ranges from over 1000 mm in the south to less than 300 mm in the north. Examples of average rainfall (1961–1990 period):

- Saint-Louis: 250 mm
- Dakar: 400 mm
- Kaolack: 550 mm
- Ziguinchor: 1200 mm

Temperatures are generally high. Minimum temperatures occur in January (15–20°C), while maximum temperatures are reached during the rainy season (35–40°C) (MEDD, 2020).

- **Vegetation Distribution**

Vegetation distribution closely follows rainfall patterns. Overall, three (3) major phytogeographic zones can be identified, arranged from north to south:

- Arid ecosystem (Sahelian zone). Rainfall: less than 600 mm Vegetation: open landscapes dominated by acacia trees and annual grasses, forming a more or less continuous ground cover.
- Semi-arid ecosystem (Sudanian zone). Rainfall: between 600 and 1000 mm.
- Vegetation: savanna woodlands, ranging from tree savannas to dry forests with a herbaceous layer dominated by perennial grasses.
- Sub-Guinean ecosystem (Guinean zone). Rainfall: over 1000 mm. Vegetation: dense semi-dry forests with two canopy levels and an undergrowth composed of woody shrubs, lianas, and grasses.
- In addition to these zonal vegetation types, azonal formations also exist, including: river and lake ecosystems, mainly in the south of the country, with gallery forests along watercourses; coastal ecosystems, with mangrove mudflats in the delta and estuarine areas of the Senegal, Saloum, and Casamance rivers, Interdunal depressions or "Niayes", located along the Senegalese coastline. (Republic of Senegal, 2019)

- **Water Resources**

With 700 km of coastline, Senegal's waters cover a continental shelf of nearly 30,000 km² (MEDD, 2020). The main rivers are:

the Gambia River (1,150 km long with a drainage basin of 77,000 km²), the Senegal River (1,770 km long with a drainage basin of 337,000 km²), the Casamance River (approximately 200 km long with a drainage basin of 21,150 km²).

Other secondary water bodies include: the Saloum River, Lake Guiers, the Anambé, and the Kayanga (MEDD, 2020).

- **Relief**

Senegal is a country with predominantly flat terrain (Gaye, 1997). Altitude is below 50 meters across nearly 75% of the territory. The highest point is located in the southeastern part of the country, in the foothills of the Fouta Djallon, reaching 581 meters. Other notable landforms include: the Thiès escarpment (128 m in the west), the Ndiass hills (104 m in the central-west), and the Mamelles hills in Dakar (105 m) (MEDD, 2020).

- **Soils**

Senegal has six natural regions. Arable land represents 19% of the country's area, or about 3.8 million hectares, distributed as follows: 57% in the Groundnut Basin, 20% in Casamance, 10% in Eastern Senegal, 8% in the Senegal River Valley, 4% in the Sylvo-pastoral zone, 1% in the Niaye (MEDD, 2020).

Soil types vary across regions:

- In the north: sandy and dry soils, mainly brown-red soils and slightly or non-leached tropical ferruginous soils;
- In the central regions: leached tropical ferruginous soils;
- In the south: lateritic soils, including ferrallitic soils and poorly developed erosion soils;
- Along the coastline: hydromorphic soils, particularly in the Niayes depressions;
- In riverine and marine areas: halomorphic (saline) soils (Op. Cit.).

- **Population and Economy**

In 2023, the population of Senegal was estimated at 18,275,743 inhabitants, including 9,184,886 women and 9,090,857 men (ANSD, 2023). The median age is 19 years, reflecting a predominance of young people, as is the case in other African countries. In 2017, the urban population had reached 46.7%. Senegal is home to more than 20 ethnic groups, though many of them are small in number (Gaye, 1997).

From an economic standpoint, Senegal relies largely on the exploitation of natural resources (mining, fishing, agriculture), as well as sectors such as construction and tourism. The Exploratory Report on the Green Economy (2019) estimates that around 60% of the population (particularly in rural areas) depends on these natural resources for their activities and/or income.

On the other hand, Senegal's trade balance remains in deficit. To meet its consumption and equipment needs, the country imports more high-priced goods than it exports in processed products (such as peanut oil, peanut cake, canned fish, green vegetables) or raw materials (cement, phosphates) (ILO, 2013). In 2022, exports amounted to 3,563 billion CFA francs (€5.4 billion), representing a 24% increase compared to 2021 (French Treasury Department, 2023). Mali is Senegal's top export destination, accounting for nearly 20% of total exports, followed by India (15.2%) and Switzerland (12%). In 2022, exports were dominated by petroleum products (18.2% of total value), followed by gold (16.3%) and phosphoric acid (13%) (French Treasury Department, 2023).

Chapter 1

Green Building in Senegal

Partial Introduction to Chapter 1

This Chapter 1 addresses the geographical stakes and characteristics of green building in Senegal, focusing on the types of housing in the country, the possible interactions between the sector and the Senegalese geographic space, as well as the mapping of key stakeholders and major achievements.

As a reminder, green building is part of the “construction” branch within the construction and public works (BTP) sub-sector of the industrial sector. According to Ndiaye (2016), the BTP sector accounts for 10.8% of Senegal's GDP and encompasses various activities, such as design, construction, deconstruction, rehabilitation, maintenance, and decontamination of public and private structures (ANSD, 2016). Construction activities include residential buildings (apartment blocks, villas, houses) as well as non-residential buildings. The latter include transport infrastructure (roads, railways, ports, airports, etc.), urban infrastructure (shopping centers, monuments, etc.), industrial structures (factories, warehouses, tanks, etc.), and hydraulic infrastructure (dams, dikes, etc.). In 2016, the "construction" sub-sector accounted for 2.6% of nominal GDP and 11.1% of the added value of the secondary sector (at current prices) (Ibid.).

1.1. Geographical Stakes of Green Building in Senegal

1.1.1. Types of Housing in Senegal

According to Craterre (2014), based on the analysis by architect Mbacké Niang, seven major housing (architectural) typologies reflecting the built environment of Senegal can be identified:

- Traditional architecture, primarily made from wood, straw, and/or raw earth (adobe or rammed earth);
- Colonial architecture, mainly made from stone, wood or metal, and fired earth;
- Modern architecture, inspired by “Sudano-Sahelian” or “international” styles, made primarily from cement blocks, reinforced concrete, and glass;

- Planned, mass-produced housing, based on standard designs of varying quality, in either colonial or modernist style;
- Regular popular housing, based on "traditional" urban planning, and either colonial or composite styles;
- Irregular popular housing, also based on "traditional" urban planning, showing a mix of all typologies and a heterogeneity of styles and materials;
- Spontaneous, precarious housing, such as slums, usually associated with unsanitary conditions.

The predominant form of housing in Senegal is single-storey houses (57.2%), with 84.9% of them financed by households themselves (ANSD, 2014). Some households still live in huts (21.9%) or shacks (2.4%) (Ibid.). The trend toward modern housing does not necessarily indicate improved living conditions. Nearly three out of ten households (29.3%) live in overcrowded conditions, with three or more people per room (30.9% in rural areas vs. 27.8% in urban areas) (ANSD, 2014). Ownership (72.6%) and renting (23.1%) are the most common forms of tenure.

Within this typology of single-storey homes, construction materials range from plant fibers used in traditional village shelters to cement block and reinforced concrete houses, often with a flat concrete roof. This latter form is particularly valued, as it allows for vertical expansion with an additional floor. However, it remains financially inaccessible to much of the population. Many buildings instead follow a construction model using cement block walls, reinforced concrete beams, and sloped sheet metal roofs (Craterre, 2014).

As for cement blocks, there are several types of production. There is vibrated cement block, produced by companies using "egg-laying" block machines to meet European quality standards, and hand-molded blocks, produced on-site near construction areas by artisans locally called moleurs (a local adaptation of the French word mouleur, meaning molder). This latter method yields a more affordable material but of lower quality.

The use of concrete comes with disadvantages: it requires large quantities of steel and formwork wood, both relatively expensive. This often results in the use of poor-quality steel and poorly cast concrete, leading to a final product with subpar quality (Craterre, 2014).

Regarding materials, some are sourced locally (cement, dune sand, gravel, etc.), while others are imported (electrical accessories, plumbing fixtures, aluminum, glass, ceramics, wood, and its derivatives). The significant share of imported products means that any increase in global market prices directly affects construction costs (ANSD, 2014).

Photo 11: House under construction

Source: Craterre, 2014. (Construction site of a single-storey house with a reinforced concrete frame and concrete blocks along the Dakar–Thiès axis. Credits: Craterre)

Photo 12: Example of Traditional Housing

Source: Craterre, 2014 (Traditional architecture, hut made of plant fibers along the Saint-Louis – Richard Toll axis)

Photo 13: Example of a Colonial Building in Rufisque (Dakar)

Source: Author, 2025

1.1.2. Green Building and the Senegalese Geographic Space: Multiple Possible Interactions

In its pursuit of modernity, African architecture has unfortunately severed ties with its centuries-old bioclimatic traditions—an ensemble of know-how that enabled inhabitants to cope with sometimes extreme climates. Today, across the continent, buildings are often poorly adapted to the hot climate, thermally uncomfortable, and energy-intensive (Lepoint, 2017).

Senegal is no exception. Across the globe, there is growing interest among the scientific community and builders in the use of natural fibers, including various types of reeds, to reduce the energy footprint of construction. In order to lessen the environmental impact of the building sector in Senegal, it is essential to produce and use new materials that consume fewer non-renewable resources during their production and implementation, and that offer better energy efficiency for buildings (especially in terms of insulation) (Craterre, 2014).

Photos 15 and 16: Old Structures Made of Raw Earth

Sources: www.orgueiletpatrimoine.fr and www.missionmali.ch, based on Legrand (2017) (left: Bam Citadel in Iran; right: Djenné Mosque in Mali – earthen structures listed as UNESCO World Heritage Sites).

According to proponents, the potential interactions between green building and Senegal's geographic space are numerous. Socially, traditional housing in various communities across the country had an ecological dimension, as it was designed in response to the local climate using readily available materials. Furthermore, the country has a significant stock of materials that can be used in the ecological construction sector.

1.1.2 A. Overview of Traditional Housing through a Few Examples

As in many parts of the world, traditional Senegalese housing adapted to the climatic and sociological constraints of each region and community[5].

5 https://www.planete-senegal.com/senegal/habitat_traditionnel_senegal.php/

- **The Diola Housing**

The Diola are a West African ethnic group settled across a territory stretching from The Gambia to southern Senegal (in Casamance) and Guinea-Bissau. Composed of several subgroups, their identity is characterized by the use of Diola languages. Their traditional houses, sometimes multi-story and with impluvium roofs, form their typical habitat[6]. These houses are mainly built with earthen walls.

Unlike other ethnic groups in Senegal, Diola houses typically have multiple rooms (an average of five), including bedrooms, a common room that can be considered a living room, and a large covered terrace that sometimes surrounds the entire house. This complex architecture also includes rice granaries, found in every home, usually located above the head of the family's bedroom.

The long rainy season typical of Casamance has influenced both the architecture and the choice of materials. All houses are built on an earthen foundation about 50 cm high to elevate the dwelling and protect it from ground moisture[7]. The roof, always rectangular, is generally covered with thatch, which prevents water from penetrating the rice granaries.

The terrace and small windows ensure good air circulation and coolness inside the house, further enhanced by the use of earth for the walls (Ibid.). Houses of the Manjaks, Mankagnes, Balantes, or Baïnouks are built on the same concept, although they may have some variations (Op. Cit.).

Photo 17: Multi-story compounds in Mlomp, southern Senegal

Source : www.planete-senegal.com/

6 https://fr.wikipedia.org/wiki/Diolas

7 https://fr.wikipedia.org/wiki/Diolas

Photo 18: Diola Hut-House

Source: www.planete-senegal.com/

Photos 19 and 20: Impluvium Houses

Source: www.planete-senegal.com/ (Impluvium huts in Affiniam, Eloubaline, Dilapao, Enampore, and Sélék, southern Senegal)

- **Peul and Toucouleur Housing**

The Peul people are spread from the north to the south of Senegal. Due in particular to the climate, their traditional dwellings vary significantly depending on the region. In southeastern Senegal, their houses were designed to shelter animals at night and during rainy days. These huts are circular and consist of a single very large room that can be up to six meters in diameter. This large size allows for interior arrangements. The thatched roof extends down to the ground, and an enclosed external corridor provides shelter for poultry or sheep at night (Ibid.).

This type of hut is found throughout the Guinean Fouta Djallon, in Fouladou (Kolda region), eastern Senegal (Tambacounda-Kédougou region), and to a lesser extent in Boundou (Malian border) (Op. Cit.).

Photo 21: Sedentary Peul Hut

Source : www.planete-senegal.com

The Peul people in the rural areas of Ferlo and the Senegal River valley are mainly herders. A large part of the village communities lead a nomadic lifestyle, which involves sometimes temporary dwellings. The scarcity of wood and other vegetation, as well as soil that is less lateritic than in the rest of the country, means these lightweight homes are made from straw, reeds, and acacia branches, which rarely withstand termites for very long.

Photo 22: Nomadic Peul Hut

Source : www.planete-senegal.com

Moreover, in the far north of the country, where rainfall is scarce and temperatures are high, traditional buildings are made of banco (rammed earth) with flat roofs without coverings and small windows. This is the Senegal River area, where the Toucouleur people are the majority.

Photo 23: Toucouleur Adobe House

Source : www.planete-senegal.com

- **The Homes of the Serer and Wolof Peoples**

In Serer areas, homes were most often square huts. The walls could be made of mud, adobe, straw (sometimes harvest residues), or palm stalks (Op. cit). Particularly well-ventilated, the interior layout is adapted to the corners. Among the Niominka Serer, although the dwellings are also composed of a single room, the huts are circular. Their roofs may be made of palm leaves, which are more common than straw in the Saloum region.

Photo 24: Serer Straw Hut

Source : www.planete-senegal.com

The Wolof Compounds in Central Senegal closely resemble the Serer huts, but their roofs, sometimes slightly higher, can be used for storing cereals (www.planete-senegal.com).

Photo 25: Wolof/Serer Huts

Source : www.planete-senegal.com

- **The Habitat of the Bassari**

Bassari huts are small and round. Particularly well-crafted, their roofs are made of short, tightly packed thatch twigs arranged neatly in layers (www.planete-senegal.com). The walls are mainly made of mud brick covered with a protective coating. However, in the mountainous area of eastern Senegal, some villages have opted for rough stone blocks. Marble, which is abundant in the Ibel region, is also sometimes used (Op. cit.). This method of construction, considered unique in Senegal and very rare in Africa, makes Bassari houses true masterpieces, with exceptionally well-regulated temperatures (Ibid.).

Photo 25: Bassari House-Hut

Source : www.planete-senegal.com

1.1.2. B. Geography of Eco-Materials in Senegal

The materials used in eco-construction are numerous. Some are raw natural resources (earth, straw, wood, etc.), while others are processed products (plastic, fabric, etc.).

- **Raw Earth**

> "Raw earth has been used worldwide as a building material for centuries. UNESCO estimates that 17% of the world's cultural heritage is built with earth, while UN-Habitat estimates that 40% of current global housing uses this material" (Legrand, 2017).

Raw earth generally refers to a base material consisting of a paste or mud containing varying amounts of clay or silt. It is characterized by four (4) fundamental properties:

- Texture, which refers to the grain size distribution, expressed as the weight percentage of different types of particles (pebbles, measured in centimeters;
- gravel, 20 mm to 5 mm; sand, 5 mm to 0.06 mm; silt, 0.06 mm to 2 µm; clay, which consists of platelets rather than grains and is smaller than 2 µm; and metallic oxides, which have coloring properties and are also smaller than 2 µm);
- Plasticity, the soil's ability to take a given shape and be deformed without breaking elastically;
- Compressibility, which defines the relative volume change of the earth under a given pressure and moisture level;
- Cohesion, or the ability of the components to stick together. This is somewhat related to plasticity (Wikipedia, 2020 and Legrand, 2017).
- To identify the different components of a soil, a sedimentation test can be carried out. Earth suitable for construction must be mineral (not organic) and contain a clay content between 15% and 35 (http://www.maison-en-terre.fr/test-sedimentation.php).

In Senegal, the soil — the surface layer of the earth — is "*mainly lateritic or, to a lesser extent, sandy*" (Craterre, 2014). Whether gravelly, clayey, or composed of lateritic crusts or hardpans, their distribution is significant across the country. Studies of soil samples show that all degrees of induration can be observed — from almost loose materials to the hardest blocks that "resist a hammer" (ILO, 2013). The most important quarries are located in eastern Senegal. The clay minerals available are classified into several groups, two of which (kaolinites and illites) contain the main components of clays used in the ceramics industry (clay bricks, tiles, and pottery) (Ibid.).

- **Straw**

Straw is defined in 38 Dictionaries and Correspondence Collections as "the dried stalk of grasses." Three (3) types of straw can be identified in Senegal:

- Grasses of the Poaceae family (also called Gramineae), mainly Andropogon gayanus, an annual plant about 80 cm tall at maturity. It is mainly harvested in eastern Senegal, a region largely covered by savannah vegetation. For reference, savannah is characterized by a continuous herbaceous layer (at least 80 cm high) and a tree or shrub layer. It covers a large portion of Senegal, about 5,077,000 hectares (MEPN, 1982). There are four types of savannah: grassy savannah, shrubby savannah, tree savannah, and wooded savannah. These areas are regularly affected by fires during the dry season (ILO, 2013). This category of "straw" also includes crop residues (millet, rice, sorghum, etc.).
- Bamboo, or Oxytenanthera abyssinica, also known locally as krinting. It is found in eastern Senegal and Casamance. Traditional partitions made from it are widely crafted and sold across the country (ILO, 2013).
- Typha, an invasive plant at the mouth of the Senegal River. It has been the subject of uprooting plans and various exploitation projects (ILO, 2013). It covers over 3 million hectares and provides 520,000 tons of dry matter per year (Craterre, 2014).

Implementation and Processing of Straw

Various techniques are used depending on the form. Dry straw can be used whole (woven for roofing, in panels or bales for partition and wall construction, which can be coated with lime or earth, or inserted into wooden, particleboard, or plaster panels), or chopped and mixed into earth to make adobe or molded bricks. Typha can be processed into typha bricks. An artisanal method has been developed in Ross-Béthio (northern Senegal) (ILO, 2013). Mixing typha with cement is also being explored for prefabricated products such as floor/roof tiles or insulation panels, in compressed or particle form. Bamboo is traditionally used for weaving canisses (krinting), usually in standard sizes of 3 x 1.5 m. Its length and good flexural strength make it ideal for roofing battens in thatched roofs (ILO, 2013).

Wood

Using wood in eco-construction can be controversial due to deforestation concerns. However, some species are promoted:

- Eucalyptus: A fast-growing tree from Australia, scientifically introduced in the 1970s. It grows in poor, dry soils and is well adapted to the Sahelian zone (ILO, 2013).
- Filao (Casuarina): Also native to Australia, it was introduced to Senegal to combat coastal erosion. It is a pioneer species capable of thriving in nutrient-poor soils (Op. Cit.).
- Neem (Margosa tree): Native to tropical Asia, it is highly resistant and well-adapted to the Sahelian climate. It is almost ubiquitous across Senegal.
- Imported wood: Includes species commonly exported from Côte d'Ivoire (the main exporter in the UEMOA zone) for their durability, affordability, and availability: Wenge, Azobé, Iroko, Padauk, Fraké, Framiré, Kosipo, Tola (ILO, 2013).

Uses in Construction

Wood can be used for roof frameworks, load-bearing wall structures, wall cladding, joinery (doors and windows), interior flooring and exterior decking (ILO, 2013)

Other Materials

Includes, among others:

- Textiles: Construction textiles must be strong, like the national malicane, a very durable raw cotton. It is used to make Mauritanian tents made of double-layered fabric (ILO, 2013). Coated fabrics, such as those with PVC, offer better architectural performance through improved durability, waterproofing, rot resistance, and UV protection (ILO, 2013).
- Plastics: Can be recycled and reshaped into various products such as plastic tar or durable pavers (by mixing liquid plastic with fine sand).
- Stones/Gravel: Sandstone quarries are located in the regions of Thiès and Tambacounda: "1,500,000 tons are produced in Toglou and 3,000,000 tons in Mbang, according to the Directorate of Mines and Geology […].

The Taïba area in Thiès has large deposits of flint" (UN-Habitat, via Craterre 2014).

- Gum Arabic: Locally called Dakaandé, it is a "solidified sap exudate obtained by tapping the trunk and branches of Acacia Senegal or A. seyal [...]. Its remarkable solubility, low viscosity in water, and emulsifying, stabilizing, thickening, and binding properties make it highly valuable" (Craterre, 2014). It can be mixed with earth to make it waterproof, and/or used as a binder or additive. It is mainly found in arid and semi-arid regions of Senegal.

Starting from the geography of materials that can be used in eco-construction, it is now relevant to discuss the characteristics of the sector in Senegal through its key players and achievements.

1.2. Characteristics of the Eco-Construction Sector in Senegal: Mapping the Main Stakeholders and Projects

1.2.1. Key Stakeholders

Based on the 2013 ILO report, stakeholders in eco-construction in Senegal can be divided into two main categories, each comprising several subgroups: the public sector (ministries, agencies, academic and professional institutions, etc.) and the private sector (NGOs, associations, cooperatives, consulting firms, companies, others). The following presentation may not be exhaustive and/or may require updates.

The Public Sector

Within this category, we can identify ministries and other state structures such as:

- The Ministry of Urban Planning, Housing, and Public Hygiene (MULHP), which oversees agencies like Société Immobilière du Cap-Vert (SICAP SA) and Agence Sénégalaise des HLM (SNHLM). These two national bodies serve the state and local communities by constructing and marketing social housing. In December 2019, the ministry launched the "100,000 Social Housing Units" project aimed at the most disadvantaged populations. The construction phase was expected to last 5 years, requiring 7 million m² of tiles, 7 million m² of local roofing tiles, 600,000 doors, 400,000 windows, and 1 million jobs for young people.[8] Thus, eco-construction materials could play a significant role in this initiative.

- The Ministry of Environment and Sustainable Development (MEDD). This ministry is responsible for forest management and exploitation, the enforcement of related regulations—particularly concerning timber—as well as managing plantations and combating bushfires. It also oversees the Ecological Monitoring Centre (CSE), whose main role is to collect, input, process, analyze, and disseminate data and information on Senegal's natural resources to support sustainable management. Additionally, it supervises the National Agency for Ecovillages (ANEV), established in 2006, which aims to promote this new model of sustainable villages.
- The Senegalese Standards Agency (ASN). Formerly known as the Senegalese Institute of Standardization, the agency became the ASN by decree 2002-746 of July 19, 2002, and operates under the Ministry of Industry and Mines. ASN's mission is to develop national standards, raise awareness, inform, and train stakeholders in Senegal—particularly industrialists, distributors, consumers, and administrative agents—on quality management approaches, methods, and procedures for implementation within production and service companies, to help define a policy for the quality of goods, services, and life.
- The Agency for Investment Promotion and Major Works (APIX). This national agency is in charge of promoting investments and major infrastructure projects. Its mission is to foster partnerships between the public and private sectors and to promote investment in Senegal. It also oversees the implementation of the state's major infrastructure projects.
- Etc.

The Private Sector

The private sector also plays an important role in the construction industry more broadly. It includes, among others:

- The Order of Architects of Senegal, based in Dakar. This is a professional platform whose goal is to improve the legal framework for architectural practice and to promote architecture within Senegalese society (ILO, 2013).

[8] http://nouvelhorizon.sn/100-000-logements-sociaux-inscrivez-vous

- The Mbacké Niang Agency for Architecture and Research (AARMN), based in Dakar and headed by Mbacké Niang, an architectural advisor to the city of Dakar specialized in energy efficiency. A former president of the Order of Senegalese Architects, a member of the ASN, and an active promoter of eco-construction, Mbacké Niang is also the inventor of building elements specifically adapted to the vernacular context, in harmony with the local climate and traditions (Ibid.).
- DK.LAB Architects: This firm is based in Dakar and uses natural materials with implementation methods tailored to each project. Their research focuses on local solutions for architecture that is integrated into its environment and that allows for controlled construction costs (ILO, 2013).
- The architectural firm Terre Neuve: This is a French architecture agency that built a low-energy consumption building within the Jean Mermoz school complex in Dakar. It also operates in the Saint-Louis region in partnership with the Senegalese association Xarito (or ASEM) in Dagana (Op. Cit.).
- The Nubian Vault Association (AVN): A French association that promotes earthen construction techniques through a regional program titled "Earthen Roofs for the Sahel." It has developed various partnerships in Senegal, such as with the Saint-Louis region, the NGO Le Partenariat, and other NGOs.
- Village Pilote: An Afro-European association that, since 1994, has been defending children's rights and empowering them through sustainable integration. It operates a reception center at Lac Rose (Senegal) featuring bioclimatic buildings, organic farming, and renewable energy initiatives.
- Worofila: A collective founded in 2016, composed of architects and engineer-entrepreneurs specializing in earthen construction. Based in both Senegal and France, its main mission is the design and construction of ecological buildings in Senegal.
- Elementerre: A construction company based in Mbour, in Gandigal, specializing in compressed earth block (CEB) construction systems.
- Craterre: A global reference in the field of earthen architecture, committed to improving and spreading knowledge and best practices internationally. In 2014, Craterre entered into a contract with the Ministry of Environment and Sustainable Development (MEDD) through the DEEC, as part of the

project "Technology Transfer: Project for the Production of Thermal Insulation Materials Based on Typha in Senegal" (PNEEB/TYPHA Project).

- TERRA technologies: An organization focused on the development of sustainable energy systems and the use of earth as a building material. To demonstrate the technical and economic feasibility of "positive energy" houses, TERRA Technologies undertook the renovation of a villa in Dakar between 2009 and 2010, transforming it into a bioclimatic house (Villa Écologique Assie Gaye).
- Etc.

From the above, it is clear that there is a wide range of stakeholders in the eco-construction sector. Some of these actors are directly involved, while others play an indirect role. To further illustrate the sector, some completed projects and achievements will be discussed in the following paragraphs.

1.2.2. Projects and Achievements

There are many eco-construction projects spread across the country. These include:

- **The ecological villa "ASSIE GAYE":**

  Built in 1997 in the Sipres II/SICA neighborhood in Sud Foire (Dakar). It was renovated between 2009 and 2010 using local construction materials, primarily raw clayey earth (banco) from Sindia. The villa is said to be powered by renewable energy sources, including a wind turbine, a photovoltaic generator, and a solar thermal collector (water heater).

Photo 26: Construction view of the "ASSIE GAYE ecological villa"

Source: https://www.habiter-autrement.org/37-ecovillage-afrique/contributions-37/30

Photo 27: Dome under construction – "ASSIE GAYE Ecological Villa"

Source: https://www.habiter-autrement.org/37-ecovillage-afrique/contributions-37/30

Photo 28: View of the completed façade - "ASSIE GAYE Ecological Villa"

Source: https://www.habiter-autrement.org/37-ecovillage-afrique/contributions-37/30

- **The Lycée Français Jean Mermoz in Dakar:**

With the Agence pour l'Enseignement Français à l'Étranger (AEFE) as the project owner, the school covers an area of 17,000 m². It was designed by Agence Terre Neuve and Adam Yédid. Winner of the AFEX Grand Prize 2012, the construction is the result of a contextual approach, drawing on local resources to develop solutions adapted to the available skills in Senegal. The master plan follows the principles of trade wind circulation. The roof provides protection, and the ventilated double walls prevent the interior surfaces from heating up. Thick walls and deep window recesses help limit direct sunlight. These are all key elements of traditional tropical architecture. The teaching buildings are arranged in narrow strips, creating shaded in-between spaces, narrow elongated patios that are open at the ends and cooled by plantings, whose shapes encourage air circulation. Given the climate, all circulation areas are outdoors, serving as living spaces in their own right, architectural walkways, and sunshades. Rocks, vegetation, and ground materials are an integral part of the project, as are the sun breakers and vegetated pergola-canopies covered with bougainvillea and other tropical species (AFEX 2012, Grand Prize).

Photo 29: View of the Lycée Français Jean Mermoz in Dakar

Source: https://afex.squarespace.com/grand-prix/2015/11/7/grand-prix-afex-2012

- **The Typha-Earth Ecopavilion of Diamniadio:**

  The project owner is the Ministry of Environment of Senegal, with the participation of Craterre for the design and the companies Elementerre and Worofila for site supervision. The Ecopavilion of Diamniadio (Dakar) was designed as part of a technology transfer project focused on producing building materials based on Typha in Senegal. This project is funded by the Global Environment Facility (GEF) and the United Nations Development Programme (UNDP). It was implemented in parallel and under the same coordination as the Greenhouse Gas Emission Reduction Program through Energy Efficiency in the Building Sector in Senegal, also funded by the same donors. The construction of this Ecopavilion serves as a demonstration space to apply the results of research on sustainable materials and building systems. It also functions as a training site (a "school construction site") for building sector stakeholders—architects, control offices, engineers, project owners, students, researchers, construction companies, banks and clients, technical and financial partners, real estate developers, etc.—and acts as a showcase for innovation in sustainable construction. It is located in the urban hub of Diamniadio, in the City of Knowledge, near the School of Architecture of Dakar, which is close to the Amadou Moctar Mbow University (https://www.construction21.org/france/articles/fr/22-l-ecopavillon-dediamniado.html).

Photo 30: External view of the Typha-Earth Ecopavilion in Diamniadio

Source: https://www.construction21.org/france/articles/fr/22-l-ecopavillon-de-diamniado.html

Photo 31: Interior view of the Typha-Earth Ecopavilion in Diamniadio

Source: https://www.construction21.org/france/articles/fr/22-l-ecopavillon-de-diamniado.html

- **The Bioclimatic Building of Alioune Diop University of Bambey (UADB):**

  Known as the PGF-Sup Building (from the Higher Education Governance and Financing Program), it was one of the six winners of the 2019 Aga Khan Award for Architecture—one of the most prestigious global architecture prizes. The originality of the Alioune Diop University building, as recognized by the 2019 Aga Khan Award jury, lies in several key elements:

- The use of bioclimatic strategies in the construction;
- A large canopy and a lattice wall that protect the building from direct sunlight while allowing for air circulation;
- The use of local construction techniques;
- Consideration of sustainability principles;
- And the aim to minimize costs and maintenance requirements.

Photo 32: Bioclimatic building of Alioune Diop University of Bambey (UADB

Source: https://www.archdaily.com/893667/lecture-building-at-the-ali

As reported: *"The construction techniques used here can serve as a prototype for other buildings. The sustainable development process behind it can serve as a model for implementing contextual, environmentally respectful architecture."* The Aga Khan Award for Architecture is the fourth prize won by this building between 2017 and 2019 (UADB official site).

- **Other achievements:**

Briefly worth mentioning are the BTC constructions of Village Pilote, the BTC train stations along the Dakar-AIBD TER line, as well as other projects by ELEMENTERRE, the Nubian vaults by AVN, etc. (more details in the second part).

Partial Conclusion of Chapter 1

Chapter 1 has presented the geographical challenges and characteristics of eco-construction in Senegal through various types of housing, potential interactions between the sector and Senegal's geographical space, as well as the mapping of key stakeholders and major projects. This overview of the first sector under study in this research now paves the way to address the second sector, namely alternative energy.

Chapter 2

Alternative Energy in Senegal

Partial Introduction to Chapter 2

This second chapter, which deals with alternative energy, aims to provide an overview of Senegal's situation in this field before presenting case studies. It is divided into two sub-chapters: the first (2.1) presents the challenges of renewable energy in Senegal, and the second (2.2) discusses the actors and types of renewable energy developed in the country.

2.1. Challenges of Renewable Energy in Senegal

Senegal has been engaged in the development of renewable energy since 1962. However, initiatives have often been pilot actions conducted in isolation (PANERs, 2015). Despite the discovery of offshore gas and oil (560 billion m^3 of gas and 530 million barrels of oil) off the Senegalese coast since 2014, the government is continuing its renewable energy development plan, which aims to gradually integrate renewable sources into the energy mix.[9]

At the inauguration of the first phase of the Taïba Ndiaye wind farm in February 2020, President Macky Sall stated: *"We are currently at approximately 22% renewable, i.e. clean energy, and by the end of the construction of this wind power plant in Taïba Ndiaye, we will reach 30%".*[10]

2.1.1. Socio-Economic Challenges

Already in 2014, the final report RGPHAE 2013 by ANSD noted that nearly 8 out of 10 households used a modern energy source for lighting (57.5% electricity, 20.4% rechargeable lamp). For cooking, wood (50.4%) and gas (32.7%) were the most used fuels (86% wood in rural areas, 58% gas in urban areas). Senegal is therefore characterized by unequal access to energy services both between and within regions. These disparities hinder balanced territorial development (Source). The country's dependence on fossil fuels significantly

[9] http://www.infos15.com/les-energies-renouvelables-et-alternatives-au-senegal-3eme-partie.html

[101010] www.lequotidien.sn/inauguration-de-la-centrale-de-taiba-ndiaye-le-senegal-dans-lere-de-leolienne/

hampers economic growth and the ability to invest in rapid access to basic social services. For example, Senegal's oil bill rose from 401 billion CFA francs in 2009 to about 708 billion in 2013 and 733 billion in 2019, causing a massive outflow of foreign currency and negatively impacting the trade balance. Additionally, 57% of the income from the country's main export products is used to pay this bill (Op. Cit.). Thus, the development of renewable energy could provide Senegal with modern, affordable, and sustainable energy services, boosting the economy to acceptable development levels that ensure improved well-being, poverty reduction, and increased energy security. It could also lead to the development of industrial sectors through private sector involvement and contribute to job creation in the energy field.

2.1.2. Political Challenges

The development of renewable energy could help reduce the risk of conflicts between states, unlike fossil fuels which often cause tensions due to their transboundary nature.[11]

2.1.3. Environmental Challenges

In Senegal, the greenhouse gas emissions profile by sector is as follows:

- 49% from the energy sector;
- 37% from agriculture;
- 12% from waste;
- 2% from industrial processes.

The public electricity production infrastructure is largely thermal (93% of installed capacity), and the dominant fuel is fuel oil (75%). This shows a strong dependency on imported petroleum products for electricity generation. Therefore, implementing a renewable energy development policy aims to:

- Reduce greenhouse gas emissions;
- Mitigate pollution (air, water, soil, and biosphere);
- Preserve ecosystems;
- Prevent accidents related to fossil fuel exploitation (Op. Cit.).

Regarding deforestation, charcoal consumption, which represents 84% of household energy use (for cooking, tea, incense, ironing, etc.), amounts to 58 kg

[11] http://www.energie.gouv.sn/enjeux-geopolitiques/

per person per year, according to FAO. It causes the annual loss of 40,000 hectares of forest. Since the 1960s, Senegal has reportedly lost 42% of its forest area.[12]

2.2. Stakeholders and Types of Renewable Energy Developed in Senegal

2.2.1. Stakeholders in the Renewable Energy Sector in Senegal

As in the eco-construction sector, stakeholders can be divided into two main categories: public sector actors (ministries, agencies, academic institutions, etc.) and private sector actors (NGOs, associations, cooperatives, companies, donors, others)

(Note: This list may not be exhaustive and could require updates.)

- **Public Sector Stakeholders**

Some of the main actors include:

- *Ministry of Petroleum and Energy (MPE)*

It designs and implements national policy in the oil and energy sectors under the guidance of the President. It oversees the promotion, exploration, and management of prospective hydrocarbon zones and ensures the relevance of technological choices for solar, hydro, and wind energy. It also ensures the valorization of scientific and technological achievements (MPE website, 2020).

- *Agency for the Economy and Energy Management (AEME)*

Created in July 2011 (Decree 2011-160), AEME serves as the operational arm for implementing national energy management policy. It promotes energy efficiency, rational energy use, and energy diversification across all sectors and consumer groups.

- *École Polytechnique of Thiès*

Launched a Master's program in Renewable Energy Engineering in 2018. In 2019, student Serigne Fallou DIOUF, with the Eco-Energy project, won third place in the 2019 Inter-University Renewable Energy and Energy Efficiency Competition (CIEREE) organized by GIZ as part of PESEREE.

[12] http://www.nebeday.org/p/charbon-de-paille.html

- *Electricity Sector Regulatory Commission (CRSE)*

An independent body under the Ministry of Energy, responsible for regulating electricity production, transport, distribution, and sales.

- *National Electricity Company (SENELEC)*

A majority state-owned public limited company that holds a monopoly on the purchase, transport, and wholesale distribution of electricity. Law 98-29 on the electricity sector gives SENELEC the responsibility to issue calls for tenders for new electricity generation capacities, under the supervision of CRSE, which ensures fairness, transparency, and non-discrimination in these procedures.

- *Etc.*

2.2.2. Types of Renewable Energy Developed in Senegal

These include solar, wind, hydroelectricity, and biomass.

2.2.2.A. Solar Energy

With 3,000 hours of sunshine per year and a global solar irradiation of 5.8 kWh/m²/day (less during the rainy season), Senegal has favorable potential for harnessing solar energy. So far, the exploitation of this resource has taken place through two sub-sectors: photovoltaic solar energy and thermal solar energy

The photovoltaic solar sub-sector is the one that has seen the most significant development. The first installations, carried out in the 1980s, continued steadily over the years through various projects and programs.[13] With the creation of ANER (National Renewable Energy Agency) in 2013, and Senegal's commitments at COP 21 in Paris in 2015, the photovoltaic sector has seen major public and private initiatives. For example, in January 2018, Senegal inaugurated its fourth solar power plant, Ten Mérina (30 MW), located in the municipality of Mérina Dakhar (Thiès region), 120 km from Dakar. This plant added to: Senergy (30 MW), located in Santhiou-Mékhé, 10 km from Ten Mérina; Senergy II (20 MW) in Bokhol (Saint-Louis); Malicounda (22 MW). [14]As of February 2020, the total reached eight (8) solar power plants, with the addition of: Kahone (near Kaolack, 20 MW); Sakal (near Louga, 20 MW); Diass and Mékhé (both in the Thiès region)[15]

[13] http://www.energie.gouv.sn/solaire-thermique

[14] https://www.jeuneafrique.com /515854/economie/le-senegal-inaugure-la-centrale-solaire-de-ten-merina/

[15] https://apanews.net/news/thies-pole-des-energies-renouvelables-au-senegal

ANER has also carried out several social projects, such as:

- Public Lighting Pilot Phase (PRODERE I), a program initiated by UEMOA in 2013 to reduce the energy deficit in member countries by 2017 through renewable energy. In Senegal, it involved supplying and installing autonomous photovoltaic solar streetlights and low-consumption lamps in: Municipal districts in the suburbs of Dakar, Some religious cities, Various towns across the country. In total, 1,835 autonomous photovoltaic solar streetlights were installed in: 12 municipalities in Dakar's suburbs, 6 other towns Religious cities (ANER, 2020).
- Solar electrification of early childhood centers, schools, health facilities, religious institutions, and security posts in Niokolo-Koba Park (ESIP I): This project aimed to supply solar-generated electricity to facilities dedicated to education, health, religion, and security in rural areas. In addition to the 94 kits initially planned, 26 additional kits were added later. Work was completed by December 31, 2016 (ANER, 2020).
- PRODERE II: As part of implementing its strategy called the Regional Initiative for Sustainable Energy (IRED), UEMOA, in collaboration with ANER and SABER, launched Phase 2 of PRODERE (ANER, 2020).

Photos 33 and 34: Solar streetlights at the UGB

Source: Author, March 2020 *(These solar streetlights are the result of the above-mentioned programs)*

For the solar thermal sub-sector, it is not as developed as photovoltaic solar energy. The first experiences took place between 1987 and 1989, driven by a joint venture company, SINAES-DAGUERRE, which marketed several hundred solar water heaters, solar dryers, and thermodynamic pumps. However, most of these technologies were not maintained or were abandoned due to a lack of maintenance services and low interest (ECREEE, 2015).

Some examples include:

- Solar water heaters: the use of hot water for sanitary purposes is not widely developed (ECREEE, 2015). However, in urban centers, hotels, and with the emergence of a middle class, some demand has emerged. In 2014, standards for solar water heaters (NS 013-012) were established by the Senegalese standards agency. Under the ORIO PROJECT, solar power supply and the distribution and installation of 72 solar water heaters at 120 health posts and huts in rural areas of the peanut basin are planned to help achieve health care objectives and reduce maternal and infant mortality (ANER, 2020).

- Solar dryers: the goal of solar drying is to improve the capacity and efficiency of traditional drying methods (fish, shellfish, and agricultural products) by: enhancing hygienic conditions, reducing drying times, improving the quality of the final product. The principle involves arranging certain materials (transparent surface, black body, and insulating body) to create a greenhouse effect inside a chamber. This greenhouse effect increases the air temperature and promotes natural air circulation. Products to be dried are laid out on racks inside the dryer. The dryers used are of the direct or mixed type (FAO). The greatest potential for this technology is at fishing docks (ECREEE, 2015).

- Solar cookers or solar stoves: these use solar energy, often at low concentration, for cooking. There are several types, including: Panel cookers, Parabolic cookers, similar to traditional fire-based cooking, Box cookers (Fall A. B., 2009).The PCSA solar cooker was designed in the 1990s by Abdoulaye Touré and is made of 80% local materials (Fall A. B., 2009). Overall, the use of solar cookers remains marginal in households, facing challenges due to traditional cooking habits which often require the constant presence of the cook (ECREEE, 2015).

- Cold production through sorption: the sorption system relies on an absorbent material (usually a liquid or solid) that captures a refrigerant gas at low pressure. A heat source (such as a solar thermal panel or a boiler) heats the absorbent fluid, causing it to release the refrigerant gas from the absorbent (or adsorbent, in the case of adsorption), thereby increasing the gas pressure. The released refrigerant gas is then condensed (usually via a heat exchanger), and then evaporated in an evaporator, where it absorbs heat from the environment (e.g., a cold room or refrigerator). This evaporation process produces cooling, which can be used to refrigerate spaces or products. In Senegal, despite significant refrigeration needs at: fishing docks, slaughterhouses. For the development of the dairy sector in the Silvopastoral zone ...this technology is virtually non-existent in the country.

2.2.2. B. Wind Energy

Senegal has wind energy potential along a 50 km wide coastal strip from Dakar to Saint-Louis, where the average annual wind speed at 10 meters height is around 4.0 m/s (source). At heights between 30 m and 40 m, measurements have shown wind speeds exceeding 6.0 m/s (Ibid.). To harness this potential, several projects and programs were implemented between 1980 and 1997, including:

- Wind turbines in Niaga Ouolof, Mboro, Louly Ngogom, and Louly Bentegné in the 1980s;
- Pumping wind turbines (FIASA project – 200 wind turbines) under the Argentinian project in 1982;
- Wind farm in Mboro (1989) with 10 turbines of 6 kVA each Pumping wind turbines (unspecified);
- Alizés-Senegal project (45 wind turbines), starting in 1997 (PANERs, 2015).

In February 2020, Senegal inaugurated the Taïba Ndiaye Wind Farm (PETN) — the largest in West Africa at the time, with a capacity of 158.7 MW, supplying electricity to 2 million people.[16]This is Senegal's first large-scale wind power project, representing a 15% increase in the country's electricity production capacity. It was designed to produce power for at least 20 years with 46 wind turbines.[17] The PETN project also includes an investment of up to USD 20 million in socio-economic projects for the benefit of Taïba Ndiaye (Ibid.). Other planned wind projects include: Saint-Louis Wind Farm (30 MW); Kayar Wind Farm (10 MW); Potou Wind Farm (10 MW); INFRACO Wind Farm (40 MW).[18]

---

[16] https://www.agenceecofin.com/eolien/2502-74178-senegal-la-centrale-eolienne-de-taiba-n-diaye-officiellement-inauguree

[17] http://www.taibaeolien.com/sn/a-propos/

[18] http://www.infos15.com/les-energies-renouvelables-et-alternatives-au-senegal-3eme-partie.html

2.2.2. C. Hydroelectricity

This energy source is mainly supplied by OMVS (Organisation pour la Mise en Valeur du fleuve Sénégal) and OMVG (Organisation pour la Mise en Valeur du fleuve Gambie). Its contribution to Senegal's total electricity production is estimated at 8% (340 Gwh.[19] Studies show the existence of several sites with a potential of around 1,400 MW on the Senegal and Gambia Rivers and their tributaries (Ibid.). In 2013, OMVS completed the construction of the Félou hydroelectric dam with a capacity of 60 MW and an annual production of 320 to 350 GWh, adding to the Manantali dam (200 MW and 800 GWh/year), operational since 2002 (ANER, 2020). On February 4, 2017, OMVG launched the Energy Project in Kaléta (Guinea), which includes: construction of a power interconnection line linking the electrical grids of the four member countries; a hydroelectric facility in Sambangalou (Senegal).

2.2.2. D. Biomass

The biomass potential in Senegal is quite significant, particularly in the forest reserves located in the East and South of the country, as highlighted in the chapter on eco-construction through eco-materials. According to the Ministry of Petroleum and Energy's website (2020): *"[...] Biomass, through fuelwood and charcoal, is the main source of energy supply for households. It accounts for 84% of their energy consumption and is also the dominant energy source in the national energy balance, representing 42%."*

Table 2: Biomass potential of Senegal

Biomass	Available biomass per year [t] DM	Theoretical total energy potential [GJ/year]	Tonnes of oil equivalent per year [toe/year]
Typha	≈ 520,000	≈ 8,870,000	≈ 211,100
Rice husk	≈ 100,000	≈ 1,370,000	≈ 32,160
Groundnut shell	≈ 150,000	≈ 2,250,000	≈ 52,820
Maize, Millet, Sorghum	≈ 1,343,000	≈ 9,900,000	≈ 232,450
Household waste	≈ 365,000	≈ 1,700,000	≈ 40,800

[19] http://www.infos15.com/les-energies-renouvelables-et-alternatives-au-senegal-3eme-partie.html

Wastewater	–	≈ 1,060,000	≈ 24,900
Animal waste	≈ 183,000	≈ 2,100,000	≈ 50,300
Slaughterhouse waste	–	≈ 33,600	≈ 790

Notes: *[t] DM: Tonnes of Dry Matter ** GJ/year: Gigajoules per year ** toe/year: Tonnes of oil equivalent per year*

Source: MVV DECON Report, November 2010, according to ANER 2020 (table compiled by the author)

Among the programs or projects in the field of biomass in Senegal, the following can be mentioned:

- The National Domestic Biogas Program (PNB) aiming to install 10,000 digesters across the country (2015–2019);
- The Typha project by GRET (2016–2018), which aimed to establish a local, sustainable, and inclusive value chain for Typha charcoal through technology transfer. Its goals were to improve energy autonomy for communities, reduce deforestation, and restore access to aquatic resources. It was initiated in the municipality of Richard Toll in northern Senegal and aimed to adapt the experience of the Typha project in Mauritania (2011–2016) to this locality.
- The Program for Economic Cooking Energy in West Africa (ProCEAO): Launched in 2012, this project under the PERACOD program (Program for the Promotion of Renewable Energies, Rural Electrification, and Sustainable Supply of Domestic Fuels) aims to promote access to improved cookstoves in four countries (Benin, Burkina Faso, Mauritania, and Senegal). In Senegal, it focused on the regions of Kaffrine, Fatick, and Diourbel and targeted 150,000 households. Its objectives were to improve access to modern, affordable, and sustainable energy services for poor populations in rural and peri-urban areas, and to ensure quality management of modern cooking technologies in West Africa (Enda Energie, 2015).

Partial Conclusion of Chapter 2

This second chapter addressed alternative energies by providing an overview of the situation in Senegal in this area. It was divided into two sub-chapters: the first presented the challenges of renewable energies in Senegal, and the second discussed the actors and types of renewable energies developed in the country. These descriptions of the alternative energy sector now allow us to move on to sustainable agriculture in Chapter 3.

Chapter 3

Sustainable Agriculture in Senegal

Partial Introduction to Chapter 3

This third chapter presents the sector of sustainable agriculture in Senegal before a detailed analysis of the case studies is conducted. It is divided into two subsections. Section 3.1 provides a general overview of the Senegalese agricultural sector through the agrarian geography of the country, the strengths and constraints of the agricultural sector, as well as a review of various agricultural policies. Section 3.2 discusses the characteristics of sustainable agriculture in Senegal, with a focus on key actors and sustainability systems.

3.1. General Overview of the Senegalese Agricultural Sector

3.1.1. Agrarian Geography of Senegal

In Senegal, 6 to 7 major agro-ecological zones can be distinguished, based on the country's biophysical and socio-economic characteristics. Although most agricultural crops are grown throughout the country, some dominate over others. The main agro-ecological zones are:

- The Senegal River Valley (irrigated rice, legumes);
- The Niayes region (80% of the country's horticultural products);
- The Groundnut Basin (groundnuts and millet);
- The sylvo-pastoral zone or Ferlo (livestock farming);
- Eastern Senegal (cereal and cotton zone);
- Casamance (rainfed rice, fruit production, cereals, and cotton in Upper Casamance) (Sources: CIAT; BFS/USAID, 2016 and Wikipedia 2020).

3.1.2. Weight and Constraints of the Agricultural Sector in Senegal

Senegalese agriculture employs nearly 70% of the active population (4.51 million people) and accounted for 17.5% of GDP in 2013. Agricultural, agri-food, and fisheries products contribute 26% of the country's exports by value (CGAAER, 2017). Farms, which are mainly family-run, often combine subsistence crops and cash crops. The export-oriented fruit and vegetable production is considered to be growing, as is the agri-business sector (Op. Cit.).

In terms of seasonality, rainfed agriculture is by far the most dominant (67.4%), followed by irrigated agriculture (7.0%) and recession agriculture (3.5%) (ANSD, 2024).

Mechanization remains low in the sector. The provisional report from the 5th General Population and Housing Census (RGPH-5) on agriculture estimates that 12.6% of farming households own motorized equipment, compared to 73.7% that own animal-drawn equipment; and 22.9% use motorized equipment, compared to 76.5% using animal-drawn equipment (ANSD, 2024).

Regarding constraints, a combination of factors — such as poor soil quality, climate risks, lack of infrastructure, and limited access to quality seeds and inputs — have led to the underdevelopment of the sector, which fails to meet the country's food needs. As a result, the country is heavily dependent on food imports, particularly rice, the staple food, whose imports account for 65% of national consumption (CIAT; BFS/USAID, 2016). Other challenges mentioned by MAER and PRACAS (2014) include: inadequate financing systems for the sector, disorganized and poorly regulated agricultural product marketing, low levels of human capital development.

3.1.3. Review of Agricultural Policies in Senegal

Since independence in 1960, Senegal has implemented various national agricultural policies and strategies, but these have not brought lasting improvements in the sector's contribution to the national economy or rural living conditions (MAER, PRACAS 2014). Several policy phases can be distinguished:

- Before 1958 (colonial period): Focus on peanut monoculture;
- 1958–1964: Policies focused on rural development and cooperative development;

- 1965–1979: Introduction of Regional Rural Development Corporations (SRDR) based on ecological zones;
- 1979–1997: Implementation of the New Agricultural Policy (NPA) and the Agricultural Structural Adjustment Program (PASA) led to accelerated liberalization of the sector;
- 1998–2012: Several strategies and programs were implemented.;
- In the early 2000s, the first Poverty Reduction Strategy Paper (PRSP 1) emphasized reducing poverty and exclusion in rural areas.;
- In 2004, the government adopted a long-term vision through the Agro-Sylvo-Pastoral Orientation Law (LOASP) 2004–2024, forming the foundation for Agro-Sylvo-Pastoral development;
- In 2008, the GOANA initiative aimed at achieving food sovereignty and increasing production;
- From 2012, especially since 2014, Senegal launched PRACAS (Program to Accelerate the Pace of Senegalese Agriculture), aligned with the National Agricultural Investment Program (PNIA) and LOASP. It also supports the Emerging Senegal Plan (PSE), particularly Axis 1: Structural Transformation of the Economy and Growth. According to the then-Minister Pape Abdoulaye Seck, the main goals were: *"To build a competitive, diversified, and sustainable agriculture, strengthening resilience, benefiting from international trade, while ensuring stable incomes for rural populations and creating many agricultural and non-agricultural jobs, particularly for youth"* (MAER, PRACAS, 2014). PRACAS Phase 1 (2014–2017) targets: rice self-sufficiency by 2017: 1.6 million tons of paddy rice, onion self-sufficiency by 2016: 350,000 tons, improved peanut production (1 million tons, and 100,000–150,000 tons for export), off-season fruits and vegetables for export (157,500 tons by 2017). In the 2018 evaluation, rice self-sufficiency wasn't achieved, but rice production reached 1 million tons in 2017, up from 4,000 tons in 2013; peanut exports reached 149,000 tons in 2017, up from 28,000 in 2013. The targets for onion and peanut were exceeded. PRACAS Phase 2 (2018–2022) aimed to: increase production and productivity of food crops, improve their economic performance, develop and consolidate industrial value chains, strengthen the fruit and vegetable export sector, ensure domestic supply of onions and potatoes (Ibid.).

3.2. Characteristics of Sustainable Agriculture in Senegal

Several actions are being undertaken or planned to promote sustainable agriculture in Senegal, including: promotion of agroecology as a farming model, promotion of green economies (e.g., biodigesters, renewable energy), creation of dedicated forage production zones, integration of farming and livestock, and use of natural fertilizers. Civil society also emphasizes: support for land access and security for women, advocacy for the recognition of peasant seeds and farmers' rights in Senegal, support from chambers of commerce for training and processing of local products.

3.2.1. Stakeholders in Sustainable Agriculture in Senegal

As in previous sections, stakeholders in sustainable agriculture in Senegal are categorized into: public and private actors including:

- Ministry of Agriculture, Food Sovereignty, and Livestock (MASAE) — the main public actor in the sector via its various departments.
- ECOWAS (Economic Community of West African States) — launches programs through the Regional Agency for Agriculture and Food (ARAA). Example: "Support Program for Agroecological Transition in West Africa" targeting 5 countries: Burkina Faso, Mali, Côte d'Ivoire, Togo, and Senegal.
- UEMOA (West African Economic and Monetary Union) — previously signed agreements with the Ministry of Agriculture to support the structuring of maize and rice sectors through the FNDASP (National Fund for Agro-Sylvo-Pastoral Development Support).
- Others...

Private actors include:

- DyTAES (Dynamic for Agroecological Transition in Senegal) — brings together key agroecology stakeholders including NGOs (Enda Pronat, Agrisud, ACF, IED Afrique...), farmer organizations (CNCR, FENAB, FPMN, Woobin...), research institutions (CIRAD, Academy of Sciences and Techniques of Senegal), international organizations (FAO Senegal), multi-stakeholder platforms (TAFAé, PNAEB, 3AO), consumer associations (CICODEV), and pioneering elected officials (e.g., network of green and ecological mayors).

- ENDA PRONAT — emerged in the early 1980s following a study by Paul Germain and Abou Thiam titled "Pesticides: A Threat to Senegal". Initially focused on awareness-raising, ENDA PRONAT evolved into research and training for producers on alternatives to chemical pesticides. In 1996, it adopted a new research and development-centered vision.
- Agrisud — an NGO committed to long-term support for Southern communities to live decently on and from their land. It offers training in agroecological principles, supports the application of best practices, and guides the economic management of production.
- Action Against Hunger (ACF) — founded in 1979, it combats global hunger. Its mission is to save lives by eliminating hunger through prevention, detection, and treatment of malnutrition, particularly during and after emergencies linked to conflict and natural disasters.
- National Federation for Organic Agriculture (FENAB): FENAB is a non-political, non-religious, and non-profit association. It was established in March 2008 and is open to all organizations of producers, sellers, processors, consumers, and support organizations involved in the development of organic agriculture in Senegal and African integration. FENAB brings together six (6) support organizations (ENDA PRONAT, AGRÉCOL AFRIQUE, GREEN SÉNÉGAL, ASPAB, GIT, and CEAS) and eighteen (18) Producer Organizations. It has set up the National Platform for Ecological Organic Agriculture, which includes over 200 organizations spread across four agro-ecological zones of the country.
- Etc.

3.2.2. Sustainability Systems in Senegal

3.2.2. A. Traditional Family Farming

> In the past, in Senegal, farming practices and cultivation methods were, as much as possible, based on the "possibilities and constraints of the environment." Generally, traditional agricultural production systems made responsible and sustainable use of natural resources (soil, vegetation, sunlight, water, nutrients, biomass, etc.) (CNCR, 2013).

Traditional family farming is considered an "old" system. According to CNCR (2013), peasant farming has long aimed to maintain a long-term balance within the overall system, by valuing production diversity, soil fertility, and sustainable land management. Observing traditional farming practices reveals a multi-layered agriculture based on synergistic interactions. This associative cropping system relies on a wide variety of plant species with different characteristics: perennial or semi-perennial, seasonal or multi-seasonal. These different plants are associated and complement each other through their morphological and physiological diversity, promoting diversified production that contributes to food self-sufficiency.

In this model, trees are fully integrated into the agricultural system and are not viewed strictly as forest elements. For example, species like Kad or Nguer from Cayor, or Dimb from Saloum, play an important role in soil fertility while offering multiple uses: human or animal food, traditional medicine, etc. *"The presence of several layers of trees adapted to ecological conditions, growing in perfect harmony with associated crops (millet, cowpea, acacia or millet, cowpea, squash, hibiscus, dimb, or other combinations), has played a fundamental role [...] in the long-term viability of peasant farming systems"* (CNCR, 2013).

Today, traditional agriculture takes on various forms. Some farmers still operate within the classic system, without machinery or chemical inputs (e.g., flood-recession farming), while others, depending on their region and available means, combine chemical inputs and small-scale machinery/animal traction. The notion of organic agriculture, independent of traditional agriculture, emerged in the early 1980s through the work of NGOs.

Today, organic agriculture, officially represented by the National Federation for Organic Agriculture (FENAB), tends to overlap with agroecology, due to actors navigating across multiple practices. As defined internationally, organic agriculture requires guarantees across the entire value chain, from seed to consumer, including certification, handling, storage, and transport of food products. However, this interpretation of organic agriculture is not entirely suited to the Senegalese context, where certification is lacking for many products. Therefore, in order to include all actors with their diverse cultural practices and territorial realities, the term "healthy and sustainable agriculture" is increasingly accepted.

According to the ILO (2013), organic products in Senegal include: cereals (millet, fonio, and sorghum are cultivated without chemical inputs but also without "organic" certification), vegetables, mango, banana, sesame, cotton.

3.2.2 C. Agroecology

NGOs, in partnership with producers' organizations, are the primary promoters of agroecology in Senegal. Their support is provided through project implementation, workshops, and publications such as guides to agroecological practices, like the one by Agrisud, translated into Wolof to help small family farms adopt these practices and reduce their vulnerability to climate change (Agrisud, 2016; Mbow, 2017).

In November 2015, Dakar hosted the first Pan-African Symposium on Agroecology, organized by FAO in collaboration with the Senegalese government. The goal was to explore agroecology's potential as a strategy for climate change adaptation, poverty reduction, and food security. Several recommendations were made to place agroecology at the heart of Africa's agricultural development, particularly for small family farms. One key recommendation was for governments to initiate an agroecological transition, as conventional agriculture has shown multiple limitations under current policy directions.

Senegal was chosen as a pilot country by FAO for a project aimed at scaling up agroecology. In the same vein, the Regional Agency for Agriculture and Food (ARAA), the technical arm of ECOWAS, launched a Support Program for the Agroecological Transition in West Africa, targeting five countries: Burkina Faso, Mali, Côte d'Ivoire, Togo, and Senegal. The objective is to "encourage the emergence, adoption, and dissemination of ecologically intensive practices in family farms, as well as management and organizational methods that support this adoption"[20]

Nationally, the Senegalese government included agroecological transition among the five major initiatives of the Priority Action Plan in the second phase of the Emerging Senegal Plan (2019–2024). However, this momentum has not yet translated into widespread on-the-ground application. There is no official agroecology department within the Ministry of Agriculture. The efforts of NGOs and other organizations remain limited compared to the scale of the agricultural sector and the vast territory of Senegal. Studies by CIRAD and its partners (see Dugué P. et al., 2015 – Cahiers d'agriculture) identified key barriers to adopting innovative practices in conservation agriculture and agroecology:

- Labor-intensiveness of some practices;

[20] www.iedafrique.org/Engager-le-Senegal-dans-une-transition-agroecologique.html

- Incompatibility of animal-drawn mechanization with intercropping or agroforestry;
- Yield improvements only visible after several years;
- Dependence on conventional production systems and input supply chains.

These findings highlight the need to change the scale of analysis and intervention—beyond individual plots to include social dynamics (individual and collective) associated with agroecological transition (Génération agroécologique, 2016).

Partial Conclusion of Chapter 3

This third chapter has presented the sustainable agriculture sector in Senegal, offering a general overview of the agricultural landscape, its challenges and constraints, and a review of agricultural policies. It then focused on the characteristics of "sustainable".

Conclusion of the First Part

This first part has revisited the key concepts addressed in the book—eco-construction, alternative energies, and sustainable agriculture—by placing them within the Senegalese context. Through this introductory analysis, the aim was to situate the study within its geographic, social, and economic realities, thereby providing a framework that aligns with local dynamics.

In this continuity, it is now relevant to move toward a more in-depth examination of the selected innovations, by analyzing them according to the zones where they are being implemented and disseminated.

Part Two

The Six Selected Ecological Innovations and Their Study Areas

Introduction to Part Two

This second part focuses on the selected ecological innovations, examined through the lens of their dissemination zones. For both pedagogical reasons and time constraints, six innovations were selected and organized into three chapters.

Chapter 4 is devoted to eco-construction, illustrated by the example of the Nubian vault promoted by AVN in northern Senegal, as well as compressed earth blocks (CEBs) through initiatives led by Village Pilote and the association Elementerre.

Chapter 5 explores renewable energies, examining the Taïba Ndiaye wind farm and the biochar production developed by the NGO Nébéday.

Chapter 6 addresses sustainable agriculture, through two areas: the use of biofertilizers, considered in general, and the development of urban micro-gardening in Dakar.

Beyond this thematic structure, the goal is to provide a technical overview of the presented innovations, with an emphasis on key elements or the least complex processes. The aim is to allow anyone, regardless of their level of expertise, to understand the different stages of development and implementation of these practices, and to grasp their basic principles.

This approach seeks to make accessible practices often perceived as technical, by highlighting their concrete impact and potential for local dissemination.

Chapter 4

Eco-construction: Nubian Vault and Compressed Earth Blocks (CEBs)

Partial Introduction to Chapter 4

This chapter focuses on two eco-construction approaches developed in Senegal: the Nubian Vault (NV), mainly implemented in the Fouta region (northern part of the country), and the use of compressed earth blocks (CEBs) through initiatives carried out by Village Pilote and the Elementerre association.

The first part of the analysis examines the Nubian Vault as promoted by the Association la Voûte Nubienne (AVN), including: an overview of the study area, the main outlines of the project in Fouta, a technical description of the construction method, and an assessment of the achievements and their main characteristics. The second part of the chapter focuses on CEB production and use. It includes: a geolocation and presentation of the key stakeholders, the CEB production chain, and a description of completed buildings and their technical specificities.

4.1. The Nubian Vault in Fouta (Northern Senegal)

> "Housing is an extremely important issue for us and for the environment, and it has been shown that living and housing conditions contribute most to the health and well-being of the population. The Nubian Vault perfectly aligns with our policy of resource management and poverty reduction, as it improves the living conditions of rural residents." Abdoulaye Baldé, former Minister of the Environment of Senegal

The Nubian Vault construction is an ancient architectural technique originating from the Upper Nile region in Nubia. It involves vaulted roofing built without formwork, using primarily raw earth (adobe or "banco"). The technique also allows for multi-story structures or the juxtaposition of several vaults. The oldest examples of Nubian vaults, dating back over 3,300 years, can still be seen at the Ramesseum in Luxor, central-eastern Egypt.[21]

[21] http://bamada.net/lutte-contre-le-changement-climatique-la-voute-nubienne-une-solution-innovante

The term "Nubian Vault" was popularized by Egyptian architect Hassan Fathy. In 1973, he collaborated with three architecture students who later founded the organization Development Workshop (DW). Since then, DW has conducted numerous training sessions on vault and dome construction in raw earth, particularly in Africa and Asia (Wikipedia, 2020). Starting in 1998, the technique was revisited by the co-founders of the Association la Voûte Nubienne (AVN), Thomas Granier, a French mason, and Séri Youlou, a Burkinabe farmer. They developed a simplified and standardized version of the method, aimed at making it easier for local populations to adopt (Fasozine, 2020).

Photo 35: Vaults of the Ramesseum in Egypt, Over 3,300 Years Old

Source : https://www.construction21.org/

4.1.1. Presentation of the Study Area and the Project

4.1.1. A. Presentation of the Study Area

Fouta is located in the northeast of Senegal, partly along the Senegal River and partly in the sylvopastoral zone. It does not correspond to an administrative region in the strict sense, but rather constitutes a historical and cultural entity. This area is primarily inhabited by the Toucouleur people. Its history is closely linked to that of the Kingdom of Fouta Toro, which was established in the Senegal River valley, at the southern border of present-day Mauritania and in the north of present-day Senegal. This kingdom succeeded the Kingdom of Tekrour, considered one of the oldest in the region (Wikipedia, 2022).

Some of the physical and socio-economic characteristics specific to this area were discussed in the first part of this study.

4.1.1. B. The VN Fouta Project

> "Of course, we're not going to give people fish but teach them how to fish... and on top of that, we'll teach them how to mend the net and sell the fish." (Thomas Granier, AVN)

In 2000, AVN launched the large-scale outreach program "For Earthen Roofs in the Sahel" to spread this architectural model throughout the Sahel region of Africa. The aim is to develop "a sustainable housing sector in West Africa, in order to provide communities with access to affordable and suitable buildings and to create green jobs" (AVN, 2021). The program started in Burkina Faso in 2000, then expanded to Mali in 2009, Senegal in 2010, and to Benin and Ghana in 2014 (Ibid.).

Its territorial deployment has gradually improved, always remaining, according to the authors, closely grounded in local realities and making use of various tested operational methods. It is now considered mature, having brought together and standardized the processes that best support the rapid development of the VN market (AVN, 2021).

In Senegal, AVN outlines the timeline of its intervention as follows:

- 2007–2010: Construction of the first VN buildings, including the Maison des Énergies in the Matam region;
- 2011: Opening of the northern regions in partnership with the NGO Le Partenariat;
- 2013: First local AVN team and expansion into the Thiès region;
- 2014: Focus on initiating public and community engagement;
- 2015: Collaborative partnerships with civil society organizations and a refugee resettlement project in Mauritania, near the Senegalese border;

- 2017: Implementation of the VN Fouta project with the NGO Le Partenariat (Website AVN, 2022). This last project is officially named "Project for the Promotion and Dissemination of the Nubian Vault Technique in Fouta (VN Fouta)". It operates in the areas of Podor, Matam, Kanel, and Ranérou. The project had a duration of four (4) years (2017-2020).

Subsidized by the Departmental Council of Yvelines (France), the general objective of this project was to: stimulate both local (Local Authorities) and external (associations from Yvelines) demand for institutional buildings in the Fouta region of Senegal, by raising awareness among these stakeholders about the Nubian Vault (VN) technique and encouraging them to direct their planned investments toward buildings using this eco-construction method. In addition, the project aims to make the Nubian Vault market in Northern Senegal self-sufficient through awareness-raising and ongoing support in terms of training and monitoring of stakeholders in the sector, as well as the formalization of emerging local enterprises (VN Fouta Transition Phase Report, 2021). The following methodological process is recommended for any entity wishing to undertake a construction project using the Nubian Vault (VN) technique:

- First, it is necessary to contact the Maison des Yvelines (MDY) after familiarizing oneself with the VN concept and the project carried out in Fouta;

- Then, a Project Intent Note (NIP) must be prepared. This document gathers the basic information related to the construction project. It is complemented by a framework note, which outlines the various steps needed to complete the VN building. Once completed, the NIP is sent to MDY, which forwards it to the VN Fouta project team. The team then prepares a feasibility note, evaluating the technical and financial aspects of the project. This document is sent to the project sponsor to help them make an informed decision.

- After confirming financial feasibility, a support agreement is signed between the VN Fouta project and the project owner. If no existing prototype plan is available for the intended building, specific architectural studies are also carried out.

- Finally, the implementation phase begins, with the launch of the tendering process, including the evaluation of bids, the awarding of the contract, and the actual start of the construction site (VN Fouta Project, 2017). As a reminder, the State of Senegal, within the framework of the Paris Agreements at COP21, has recognized earthen architecture as a model for economic development (Le Partenariat, 2020).

4.1.2. Technical Overview of AVN Construction

The following figure summarize the architectural diagrams:

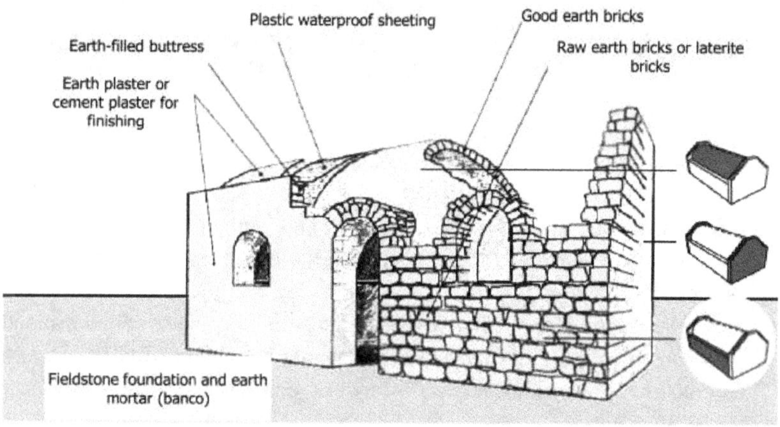

Fig. 5. Architectural Summary of the Nubian Vault (VN) by AVN
Source: AVN, 2013 (translation by Author, October 2025)

As with any construction, foundations are the first step. They are generally built using stone, although rammed earth (pisé) can also be used. The depth of the foundations depends on the nature of the soil. The structure of the building is almost entirely made of raw earth, which is moistened, kneaded, and then molded into bricks (approximately 38 × 18 × 15 cm), which are then sun-dried (adobes), or used directly in the form of mortar.

Photo 36: Raw earth, stone foundation, and load-bearing walls

Source: AVN, 2014

Walls are built with standardized thicknesses, defined according to their intended use. The same applies to openings, whether doors, windows, or built-in storage spaces. The bricks used for the vault and vault ends (areas around doors and windows) are smaller in size (24 × 12 × 4 or 5 cm) and made from high-quality earth. A cable-compass is used to precisely define the curve of the vault and serve as a guide for the mason during construction. The vaulted roofs allow for the construction of flat terrace roofs, thanks to their load-bearing capacity. Their construction does not require formwork, but the maximum width must not exceed 3.30 meters.

Photo 37: Vault ends, vaults, and buttresses

Source: AVN, 2014

After the vault has been completed and loaded, a plastic tarp is integrated into the roofing as a finishing layer to provide additional waterproofing. This is then covered with a minimum 5 cm layer of earth, designed to receive either: an enriched earth plaster (based on traditional know-how with suitable additives), or a bitumen-based coating. At this stage, parapets (acrotères) are also built using rows of masonry bricks. It's also possible to add an additional upper floor (R+1) (i.e., ground floor plus one story) (Ibid.).

Photo 38: Plastic tarp, terrace with bitumen coating, parapets

Source: AVN, 2014

Finishing plasters are applied both inside and outside the buildings. The basic or traditional finishes are mainly fine-earth plasters. A limewash, applied like paint, can be used to cover the interior walls. Cement-based plasters can also be applied to the interior walls (except the vaults) to improve impact resistance or to protect from humidity in "wet rooms" like bathrooms and kitchens. Floors can be finished with cement screeds, designed, for example, to support tiling. Cement plasters can be applied to the walls using two main techniques:

- By building external walls with bricks embedded with pebbles (called BBSC, for Briques de Banco Surface Cailloux), which create a rough surface that improves adhesion of the cement plaster (as raw earth does not allow for good adhesion);

- By attaching a wire mesh to the walls, which serves as reinforcement for the cement plaster and prevents it from peeling off.

The secondary works include interior and exterior joinery, electrical installations, plumbing, screeds, tiling, etc. (AVN, 2024).

Photo 39: Interior finishes and Secondary Works

Source: AVN, 2014

Photo 40: Summary of the AVN design and construction steps

Source: AVN, 2021

4.1.3. Achievements and Their Characteristics

According to data updated in August 2020 and published on AVN's website (2022), the AVN program in the Sahel has generated many visible and measurable impacts that address the needs of Sahelian populations. AVN manages the program through a robust monitoring and evaluation system. Data is collected from regional to national and international levels, enabling the observation of market trends, visualization of program impacts, and anticipation of areas where further reinforcement is needed.

Across all areas of intervention in Africa, the following results have been recorded:

- 44,700 beneficiaries: people living or working in Nubian Vault buildings;
- 83% of the clientele comes from AVN's core target group;
- 1,370 localities with at least one VN building;
- 4,200 construction sites completed since the program's inception.[22]

In Senegal, the VN building stock consists of approximately 100 construction sites, totaling around 9,000 m², benefiting more than 4,300 people. During the 2019–2020 season, the following achievements were recorded:

[22] https://www.lavoutenubienne.org/-resultats-et-impacts-

- 8 construction sites, totaling 714 m² of built area;
- 7 community buildings, including 3 with productive uses;
- 71% of the buildings were constructed in rural areas;
- 6 new localities received at least one VN building (bringing the total to 76 localities since the beginning of the program);
- 38% of the construction sites were sourced directly by masons, without AVN's intervention;
- 360 new beneficiaries are now living or working in VN buildings.

Regarding the VN Fouta project, the following table summarizes achievements since 2017. It should be noted that 2021 was not part of the initial project timeline but can be considered an extension of the program.

Photos 41 & 42: VN construction site, Podor Department Council (2022)

Source: Tapha SENE, VN mason, 2022

After the official end of the VN Fouta project in 2020, a transition phase occurred in 2021, leading to the launch of a new project: "DEFI-VN" (2022–2025)[23]. This project continues the work of the original VN Fouta program with a new general objective:

23 Title: Development of a Sustainable Construction and Training Sector through the Nubian Vault Technique in the Departments of Kanel, Matam, and Podor.

"By 2030, contribute to the reduction of greenhouse gas emissions by placing special attention on energy-efficient buildings in Senegal, particularly in the Departments of Kanel, Matam, and Podor." (DEFI-VN Project Presentation, 2022). The specific objective was: *"To promote and disseminate low-carbon building construction using the Nubian Vault technique in Senegal, leveraging momentum generated in Kanel, Matam, and Podor."* (Ibid.)

4.2. BTC by Village Pilote and Elementerre

The Compressed Earth Block (CEB or BTC – Brique de Terre Comprimée) is a modern evolution of molded earth blocks (adobes). The concept of compacting adobe for better quality and strength is old — early BTCs were made using wooden pestles. The first compacting machines were imagined in the 17th century, but it wasn't until the 20th century that mechanical presses appeared. The real breakthrough came in 1952, with the invention of the CINVA-RAM press by Colombian engineer Raul Ramirez, from the CINVA Center in Bogotá. The 1970s and 1980s saw the emergence of a new generation of manual, mechanical, and motorized presses, giving birth to a genuine BTC market (Guillaud et al., 1995).

According to its proponents, BTC offers several major advantages:

- Retains the ecological qualities of raw earth, while offering improved strength and quality thanks to mechanical pressing, available in various designs.
- Uniformity of production helps improve social acceptance of BTC as a modern building material.
- BTC production often includes quality control processes, allowing compliance with standards and codes, a key condition for urban use.
- In contexts where small masonry units (like fired bricks or sand-cement blocks) are common, BTC is easily integrated and serves as a complementary technology.
- BTC is highly adaptable to both rural and urban environments, and to artisanal or industrial-scale production — a major advantage for attracting policymakers, investors, and entrepreneurs.
- Water-saving during molding and reduced finishing needs (plaster and paint may be optional) are also key advantages (Guillaud et al., 1995).

Photos 43 & 44: Examples of BTC construction

Source: https://search.brave.com/images?q=image+btc+construction&img=115

4.2.1. Geolocation and Overview of Studied Actors

Village Pilote

> "All our actions are designed and implemented with the utmost respect for the environment. At our Village Pilote welcome center at Lac Rose (VPLR), Senegal, our buildings are bioclimatic, and we develop organic agriculture and renewable energy, aiming for self-sufficiency. We share this ambition with all our partners." (Village Pilote Website, 2022)

Village Pilote is an Afro-European association that, since 1994, has been defending children's rights and empowering them to uphold those rights. Its head office is in Yoff, Dakar, but it operates a welcome center in Deni Biram Ndao, a locality in the northeast outskirts of Dakar, near the Lac Rose area.

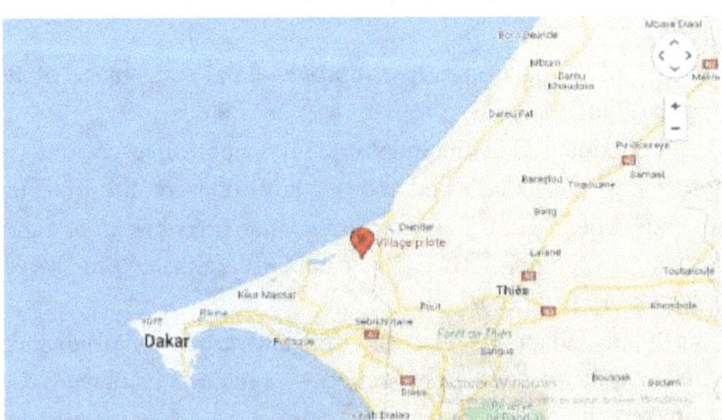

Fig. 6. Location of the Lac Rose welcome center of Village Pilote
Source: Google Maps, 2022

Elementerre

"Founded in 2010, ELEMENTERRE is a construction company specialized in earth-based building systems. Production takes place in Gandigal, 7 km from Mbour on the national road" (Elementerre Website, 2022). The administrative offices are located in Dakar, in the Fann Hock neighborhood.

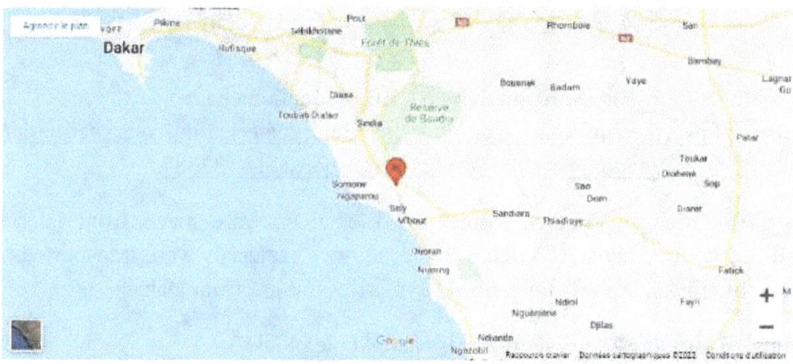

Fig. 7. Location of Elementerre's production site
Source: Google Maps, 2022

4.2.2. Compressed Earth Block (CEB) Production Chain

To manufacture Compressed Earth Blocks (CEB), laterite or clay is used, which is typically found around 50 cm beneath the surface (this may vary by location). The laterite or clay is dried by the wind and sun if it's too moist, then pulverized (sifted), moistened, and mixed with sand (in some cases). Cement or lime may be added to stabilize the mix (approximately 10% of the volume). A compression machine is then used to form uniform earth blocks, which are ready for use after a few weeks of drying (for example, 7 days under a tarp, followed by 7 days in open air).

Fig. 8. Example of a CEB production chain
Source: https://theconversation.com/le-plus-vieux-materiau-de-construction-au-monde-est-aussi-le-plus-ecoresponsable-133587

The CEBs used by Village Pilote and Elementerre are made from laterite stabilized with cement. At Village Pilote, the raw materials were transported to the site by trucks, often from the Sindia quarries (60 km from Dakar).

Photos 45 and 46: Example of a press and whole blocks

Source: Google/BTC/images

There are different types of blocks, each serving a specific function in construction:

- Whole block (used for wall filling or finishing work);
- Perforated block (has a hole in the center, used for running electrical wires or plumbing pipes);
- Semi-rounded or double-rounded block (used for constructing pillars or corners);
- U-shaped block (used for reinforcement; can contain iron bars or concrete during construction);

- ¾ block (used for constructing T- or X-shaped walls (https://www.senegal-export.com/bloc-de-terre-comprime-innovation-btp-au-village-pilote.html).

Photo 47: CEB bricks at Village Pilote

Source: Author, May 2022

4.2.3. Completed Projects and Their Characteristics

At Village Pilote, one notable example is the reception center at Lac Rose, visited in May 2022. It consists of several hybrid buildings (cement–CEB), with CEB being the dominant material, used with concrete foundations and reinforcements.

As for Elementerre, the company did not respond to the questionnaire that was sent. However, several of its projects are listed on its website, such as:

- Hotel Le Djolof in Dakar, near the small port of Soumbedioune;
- Headquarters of Le Partenariat in Saint-Louis;
- A villa in Bandia;
- The KEUR LAZ cultural center in Saly;
- The MAISON DES ENFANTS (Children's House) in Dakar.

The following photos serve as illustrations.

Photos 48 and 49: Examples of Elementerre's CEB constructions

Source: Elementerre Website, 2022 (The first photo shows a villa in Bandia (Mbour Department); the second, the Children's House in Dakar.)

Photos 50 and 51: Example constructions at Village Pilote

Source: Author, May 2022

Partial Conclusion of Chapter 4

This chapter presented two eco-construction approaches implemented in Senegal. The first is the Nubian vault, primarily developed in the Fouta region with support from the La Voûte Nubienne Association (AVN). It was discussed through a description of the intervention area, main project outlines, technical aspects of the method, and a review of completed buildings.

The second part focused on the use of Compressed Earth Blocks (CEB) through initiatives by Village Pilote and the Elementerre association. After introducing the stakeholders and sites involved, the chapter described the CEB production chain and reviewed completed buildings along with their technical characteristics.

Chapter 5

Alternative Energies: Wind Farm and Biochar

Partial Introduction to Chapter 5

This chapter is devoted to the study of two major alternative energy initiatives in Senegal: the Taïba Ndiaye wind farm and the biochar project led by the NGO Nebeday. Although different in nature and implementation, both projects align with a logic of energy transition and local resource valorization, in response to the country's environmental and social challenges.

First, the analysis focuses on the Taïba Ndiaye wind farm, the largest in the country, presenting the study area, technical features of the farm, and an assessment of its impacts. The second part is dedicated to biochar developed by Nebeday, a biomass-based alternative fuel. This includes an overview of the local context, a description of the production method, and a look at its socio-environmental effects.

5.1. The Taïba Ndiaye Wind Farm

A wind farm is a group of wind turbines located in a windy area, arranged to produce electricity. The turbines are connected to the grid at a single point.[24] Historically, windmills are the ancestors of wind turbines. They were first built in Persia in the year 620. The modern wind turbine, which converts kinetic wind energy into mechanical energy, was developed in 1890 in Denmark by Poul La Cour. After a period of decline, it saw renewed interest in the 1970s following the first oil crisis.[25]

---

[24] https://www.actu-environnement.com/ae/dictionnaire_environnement/definition /parc_eolien.php4
[25] https://www.topito.com/top-parcs-eoliens-monde

Table 4: Largest Wind Farms in the World

No.	Name	Characteristics
1	Jiuquan Wind Power Base (China)	Planned to host 7000 turbines producing 200 GW.
2	Jaisalmer Wind Farm (India)	1600 MW with a mix of old and new turbines.
3	Alta Wind Energy Center (USA)	Produces 1548 MW, in California, operating since 2011.
4	Muppandal Wind Farm (India)	1500 MW on uncultivable but windy land.
5	Shepherds Flat (USA)	Near Arlington, Oregon; 845 MW.
6	Roscoe Wind Farm (USA)	In Texas; 781.5 MW with 627 turbines.
7	Horse Hollow (USA)	Also in Texas; 735.5 MW for 180,000 homes.
8	Capricorn Ridge (USA)	662.5 MW, powers over 220,000 homes.
9	Walney Extension (UK)	Offshore; 659 MW for 600,000 homes.
10	London Array (UK)	Largest offshore farm by area, 175 turbines.

5.1.1. Study Area and Project Overview

- **Study Area**

Taïba Ndiaye is a commune in western Senegal, about 100 km from Dakar. It is part of the Méouane district, in the Tivaouane department of the Thiès region (Wikipedia, 2022).

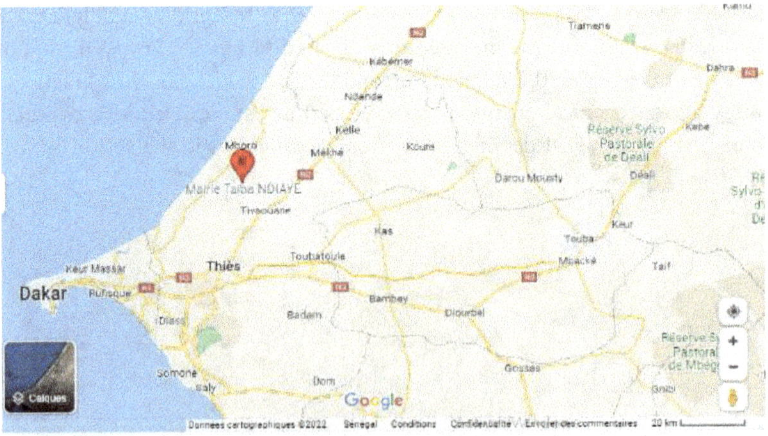

Fig. 9. Location of Taïba Ndiaye
Source: Google Maps, 2022

- **The Taïba Ndiaye Wind Farm Project**

The Taïba Ndiaye Wind Farm is the first industrial-scale wind project in Senegal. The idea originated in 2007, with construction beginning in September 2018 and full commissioning completed in July 2020. As of 2022, it is one of Africa's largest wind farms (the biggest in West Africa), with a capacity of 158.7 MW, increasing Senegal's electricity production by 15%, serving 2 million people (PETN Website, 2022).

The project's total investment is estimated at 200 billion CFA francs (342 million euros) and was built by Lekela, a subsidiary of Mainstream Renewable Power. It is managed by the Taïba Ndiaye Wind Farm Company (PETN).

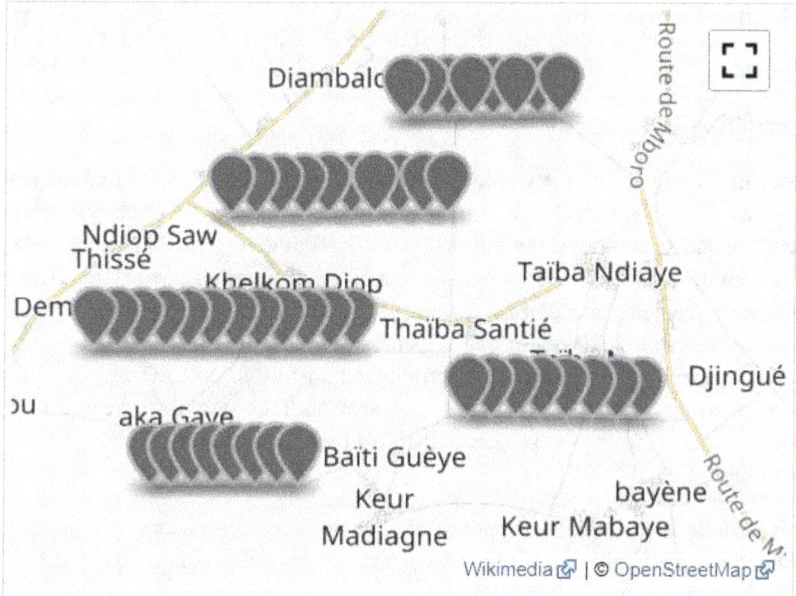

Fig. 10. Layout of the PETN wind turbines

Source: Wikimedia 2022

The wind turbines of Taiba Ndiaye are arranged in rows on different sites.

5.1.2. Some Technical Aspects of a Wind Farm

The operating principle of wind energy is relatively simple: the wind turns the blades, which in turn rotate the wind turbine's generator. The generator then converts the mechanical energy from the wind into electrical energy, known as wind power. This wind-generated electricity is directed either into the power grid or into wind energy storage batteries.

The PETN (Parc Éolien Taiba Ndiaye) includes 46 Vestas wind turbines reaching heights of 180 meters (with 60-meter blades), and capable of producing 158.7 MW of electricity. Shipped from Denmark by boat, the wind turbines were assembled in Senegal. The park covers an area of 40 hectares — the equivalent of 60 football fields (PETN website, 2022). One of the park's major advantages is its proximity to the Tobene Power plant (the second-largest independent power producer in Senegal), which allows electricity to be fed into the national grid quickly and at a reduced cost (Op. Cit.).

5.1.3. Project Impacts

According to the interview guide received on July 11, 2022, the 35 km of roads within the park have been opened to local communities, which has greatly contributed to the development of economic activities in the area, particularly agriculture. In terms of Senegal's energy consumption, Mr. Pape Biteye, Director General of SENELEC under President Macky Sall's administration, highlighted in a 2020 report that the energy produced by the park is sold up to 50% cheaper to the national electricity company (SENELEC), saving it an estimated 21 billion CFA francs per year in fuel purchase costs. On the environmental front, the project helps avoid the emission of 300,000 tonnes of CO_2 per year.

In addition to its national impact on renewable energy production and providing electricity for 2 million people, various articles have also pointed out that PETN plans to invest up to USD 20 million in socio-economic projects benefiting the Taiba Ndiaye area. These projects focus on creating sustainable livelihoods, including improving local agriculture, offering professional training and opportunities to young people in the community, and upgrading local infrastructure (PETN website, 2022). Already, the number of job opportunities during the construction phase was estimated at 600. Although it's unclear whether this number was actually reached, the interview guide mentions a skills transfer process related to the operation and maintenance of the turbines, which now involves a team of 16 Senegalese nationals responsible for all on-site activities. These individuals received training at Vestas centers in Europe and Morocco, in addition to receiving both on-site and remote assistance during the wind farm's construction phase (Interview Guide, 2022).

Other local projects have also been launched:

- Reconstruction of the Mbayenne 3 market, initiated following discussions with the Women's Association of Taïba N'Diaye. Completed in 2022, the project provided the town with a modern, clean, and covered space, offering better working conditions for vendors — especially for the sale of fresh produce, according to the project website. This initiative also reportedly improved the well-being of both sellers and their customers. A second project to renovate the central market of Taïba N'Diaye was also being prepared, according to the same source.
- The Computer Center for Students of Taïba N'Diaye, created in response to the lack of access to basic digital skills among students at the local high school — in areas such as typing, programming, and the use of digital tools. According to the project website, the center is equipped with modern and high-performance computers, and will give over 1,500 students the opportunity to familiarize themselves with digital technologies — a key driver for academic success and future employability in an increasingly digital world. This initiative is part of a broader effort to bridge the digital divide and promote equal opportunities through education, according to the project leaders.

Photos 52 and 53: Views of the PETN wind turbines
Source: Author, May 2022 (The red-colored roads are laterite)

5.2. Nebeday's Biochar

Biochar, also known as green charcoal or ecological charcoal, is a carbonized product obtained by heating biomass (of plant or animal origin) "in an environment where oxygen is limited or nearly absent" (Pthorticulture, 2020). The term also includes briquettes made from charcoal powder (from storage deposits, for example), as well as non-carbonized ones. Biochar is a fuel that, according to its proponents, can serve the same purposes as traditional charcoal.

According to wiki.lowtechlab.org, biochar produced from agricultural waste (such as straw, peanut shells, or dry bush grass) can replace traditional charcoal. Environmentally, since bush straw and agricultural waste are renewable biomass, their use reduces the risk of bushfires and helps preserve forests and biodiversity. Finally, since straw-based charcoal can be used in the same way as wood charcoal, it fits well with local culinary traditions — making it easier for communities to adopt (wiki.lowtechlab.org, 2020).

Airex Energy states that biochar has an energy content comparable to that of coal, enabling a power plant to operate at full capacity. It can either be co-fired with coal or completely replace it, effectively turning a coal-powered thermal plant into a 100% renewable energy plant (airex-energy.com, 2020). The website www.laterre.ca highlights biochar's positive effects on agricultural yields and costs. Thanks to its high carbon content, "biochar promotes plant growth and increases yields by retaining nutrients and water reserves," according to Airex Energy. The general manager also notes that several scientific studies, including those by the International Biochar Initiative, confirm its effectiveness.

Similarly, www.refertil.info notes that plant-based biochar (PBC) serves as a soil enhancer, while animal bone biochar (ABC) is used as an organic fertilizer — or both. When correctly produced, biochar has the potential to restore the natural balance of soil and improve crop yields by retaining water and nutrients, increasing drought tolerance, and enhancing soil fertility — especially for economically important crops.

5.2.1. Presentation of the Study Area and the Biochar Involved

• The Study Area

Toubacouta (also spelled Toubakouta) is a locality situated in the Sine-Saloum region, around forty kilometers south of Foundiougne, between Sokone and Karang Poste. It lies on the banks of the Bandiala, one of the branches of the

Saloum River. The village has around 3,000 inhabitants, primarily Mandinka and Serer farmers.

Tourism is developing in the area thanks to opportunities for fishing in the bolongs, hunting in the extensive surrounding zones, and wildlife excursions. Mangrove honey is also produced locally (Wikipedia, 2022).

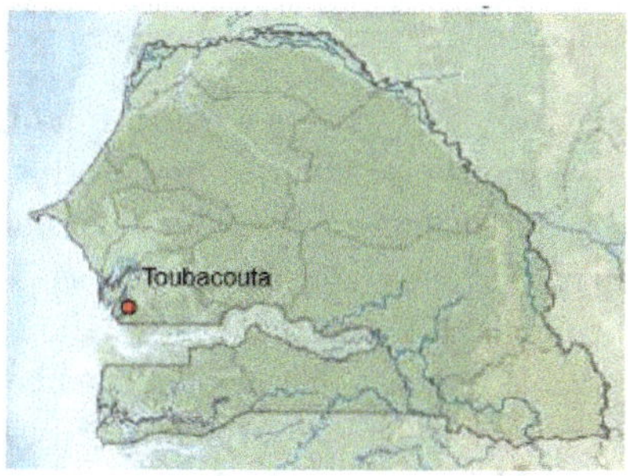

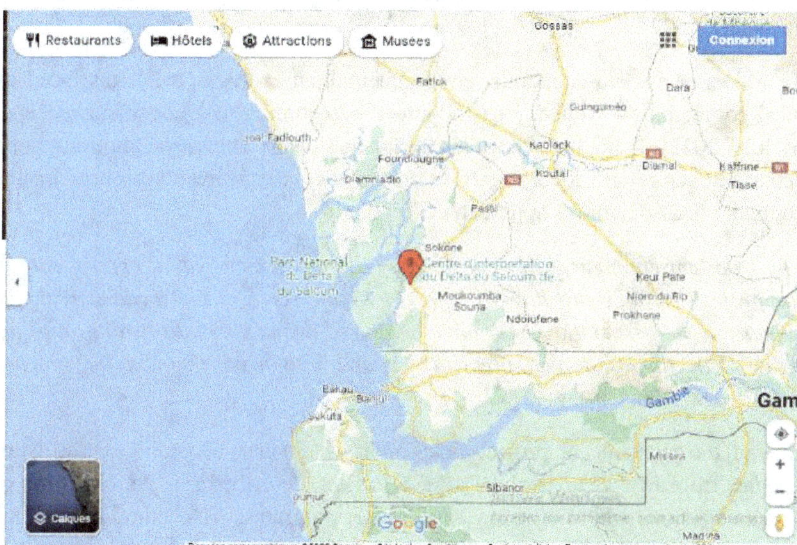

Fig. 11 & 12. Location of Toubacouta

Sources: Wikipedia and Google Maps (adapted by the author, May 2022)

• The Actor and the Biochar Implemented

Named after the Moringa Oleifera tree in Wolof, Nebeday is an association focused on the protection, management, and enhancement of natural resources by and for local African communities. Its headquarters is in Dakar, with offices in Toubacouta (Saloum region), Kaolack, and Palmarin (Nebeday Website, 2022). The association is structured into several divisions that work across multiple sectors: reforestation, agroforestry, alternative energy, etc.

Regarding its biochar work, this is not a one-off project, but an ongoing initiative that is part of various projects over time, with support from multiple donors. Nebeday began producing biochar in 2017, aiming to contribute to climate change adaptation, reduce greenhouse gas emissions, improve local living conditions in the Fatick, Kaolack, and Dakar regions. The straw collected from forests is a renewable biomass that helps reduce both wood cutting and bushfires — two major sources of emissions. Additionally, fewer bushfires allow young trees to grow, which increases carbon storage (Nebeday Biochar Factsheet, 2022).

To address the socio-economic, environmental, and energy challenges faced by local communities, Nebeday has adopted — according to a document received in July 2022 — an integrated territorial approach. The association supports economic actors and local organizations, particularly women's groups. In this context, the association's approach is described as:

"[...] not entrepreneurial, but community-based. Women play a central role in the project, as they are directly affected by the issue of fuel access. With the support of the association, they have organized themselves into rural groups for straw collection and dust preparation, and into urban groups for biochar marketing." (Op. Cit.)

One of the most recent projects incorporating biochar, according to the latest updates from July 2022, is the TEFESS Project, described as: "An extension of previous actions and integration into broader programs." (Nebeday, 2020). This project aligns with activities from PCCI2: "Sine Saloum Communities Committed to Climate Resilience" (2020–2023), and stems from regional and international cooperation. The project's total cost is estimated at €775,924.13, with the main donor being the Ministry of International Development of the European Union (Ibid.).

5.2.2. Biochar Production Technique Used by Nebeday

In general, biochar production can be carried out in different ways — from a simple hole in the ground to a sophisticated industrial pyrolysis furnace. Pyrolysis is the thermochemical decomposition of organic matter at high temperatures. The process is used to produce charcoal, activated carbon, methanol, and other chemical products from wood. The same process also applies to converting ethylene dichloride into vinyl chloride (for PVC production), making coke from coal, and converting biomass into syngas and biochar (Pthorticulture, 2020).

As for Nebeday's method, the production process involves several stakeholders engaged in different parts of the value chain. Upstream, in around 15 villages, women are organized into Women's Promotion Groups (GPFs). They handle the collection of straw near cultivated fields, livestock trails, and in the forest. Transportation is done on foot or using a cart (Op. Cit.).

Once collected, the straw is stored, then burned in metal drums using pyrolysis. These drums are slightly modified, with air vents at the bottom and a lid on top.

During ignition, the lid is left open, and the setup functions like a chimney, allowing the biomass to combust and reach high temperatures. However, the air vents are not sufficient to ensure complete combustion of the gases produced inside the drum. These gases rise and ignite upon contact with oxygen outside, resulting in secondary combustion. When the material is fully carbonized, the air intake at the bottom is blocked using sand and earth. At this stage, variability in burn times has been observed — some batches take longer than others. Once the flames are extinguished, the lid is sealed tightly, and covered with sand and earth. The cooling phase then begins, lasting about 10 minutes, to make the drum safe to handle.

After cooling, the contents are poured into a basin containing water mixed with a binder, which varies depending on the final product desired.

In an interview in July 2022, Mr. Hubert Dieme, one of Nebeday's managers, explained that the association had experimented with several binders (clay, rice starch, gum arabic) to achieve a product that was affordable and of better quality. However, due to certain constraints and challenges, clay was ultimately chosen as the main binder.

So, after mixing the charcoal dust with clay, the new mixture is dried and then compressed using a press to form briquettes. After a final drying phase, the briquettes become the finished biochar product. It's also worth noting that Nebeday purchases the charcoal dust (poussier) by kilogram from the women before the briquettes are manufactured.

Photo 54: Illustration of Nebeday's Biochar Production Method

Source: Nebeday website and files received in July 2022

5.2.3. Project Impacts

Historically, the association produced two types of biochar: Kérin Naatangué Tiger (using rice starch as a binder) and Kérin Naatangué Thiouraye (using clay).

The first type was abandoned due to a socio-cultural issue: women were uncomfortable with the idea of turning rice — a staple food — into powder to make charcoal, especially while some people couldn't afford rice to eat. Kérin Naatangué Tiger was designed to replace traditional cooking charcoal.

Kérin Naatangué Thiouraye, with its slow, smoke-free combustion, was mainly intended for use as incense (Thiouraye in Wolof), for tea-making (Ataya), and ironing.

Special kraft paper packaging was ordered from France to package the product for better visibility and marketability in supermarkets, especially in Dakar. Previously, the association bought charcoal dust from "carbonisatrices" (women's groups that carried out the carbonization) at 100 FCFA/kg for both types. However, due to some groups adding excess clay to increase weight, Nebeday decided to take over the remaining manufacturing steps in-house, under stricter quality control.

As of 2022, the recorded groups of carbonisatrices included: Lambaye, Diabala, Ndiaye Ndiaye Serere, Yerwago, Dagassoukoum, Keur Mor Diop 1, Keur Mor Diop 2, Guede, and Diokoul. Between July 2017 and June 2018, the total charcoal dust produced was 2,103.5 kg, resulting in 11,749 briquettes. The revenue earned by women from selling this dust was estimated at 210,350 FCFA (Nebeday Files, 2022). Between February 13, 2020, and June 15, 2022, production reached 22,008.5 kg, for a total of 2,208,850 FCFA earned (Op. Cit.). From January 18, 2019, to June 16, 2022, the total number of Kérin Naatangué Thiouraye briquettes produced was 65,576, weighing 16,394 kg. Each 1.5 kg bag was sold for 500 FCFA (Ibid.). In addition to the packaged product, a portion of the production is also sold in bulk.

A whole economic value chain has emerged around biochar production — with women's groups at the upstream end, and others involved downstream, such as metalworkers, artisans, transporters, etc. On top of that, the project continues to produce positive ecological impacts by preserving forest resources and improving the carbon balance.

Photo 55: Bag of Nebeday Biochar

Source: Mr. Hubert Dieme (Nebeday), received in August 2022

Partial Conclusion of Chapter 5

This chapter provided an opportunity to examine two alternative energy initiatives in Senegal: the Taïba Ndiaye wind farm and the biochar project by the NGO Nebeday.

Through analysis of their implementation contexts, technical characteristics, and impacts, these two examples illustrate different yet complementary approaches to energy transition: the wind farm represents a large-scale investment strategy with integration into the national grid. The biochar initiative, in contrast, is based on local dynamics and the valorization of available natural resources.

Chapter 6

Sustainable Agriculture: Biofertilizers and Urban Micro-Gardening

Partial Introduction to Chapter 6

This Chapter 6 focuses on two selected innovations in the field of sustainable agriculture. It begins, in Section 6.1, with biofertilizers, analyzing their agronomic benefits, the potential of the global market, the technical elements required for their production, as well as selected case studies and their impacts. Section 6.2 is dedicated to urban micro-gardening in Dakar, with a presentation of the study area, technical aspects of micro-gardens, and the observed effects of the project in the local context.

6.1. Biofertilizers

(1) One kilogram of soil can contain:
1,000 billion bacteria
1,000 billion protozoa
500 million algae
100 million fungi
Thousands of mites, nematodes, and mini-insects
10 earthworms
Insect larvae, ants, etc.

Source: Liette Lambert, "Les biostimulants, parlons-en", excerpt from André Carrier's conference, MAPAQ, March 13, 2014, Lac St-Jean, France

A biofertilizer (or biostimulant) is a biological or organic fertilizer, as opposed to chemical or mineral fertilizers. Its particularity lies in its composition: it contains nutrients and living microorganisms that contribute to improved plant growth. It enhances soil functions and fertility through the action of these microorganisms (Traoré, 2017). Biofertilizers can be liquid or solid, depending on the ingredients used in their preparation. They can also play a role in stimulating the plant's natural defenses and protecting against diseases and pests (Bio Savane, 2015).

For plant development, light, water, carbon dioxide, oxygen, and mineral elements are needed. Using their root systems, plants draw these resources from the soil for nutrition. But soil is more than a "nutrient reservoir"; it is a complex ecosystem. Even with a strong "nutritional capital", some nutrients may be immobilized and therefore unavailable to plants. This is where soil microorganisms come into play. They enable mechanisms that improve nutrient bioavailability, thus promoting plant development. So when you apply fertilizers or water, these biofertilizers or biostimulants help the plant to better assimilate those inputs (RFI, 2017). Types of Biofertilizers:

- Biofertilizers based on nitrogen-fixing bacteria;
- Biofertilizers based on symbiotic bacteria;
- Biofertilizers based on non-symbiotic bacteria;
- Biofertilizers based on mineral-solubilizing bacteria;
- Biofertilizers based on plant growth-promoting rhizobacteria (PGPR);
- Biofertilizers based on arbuscular mycorrhizal fungi (Alamri, 2016).

6.1.1. Agronomic Benefits and Global Market Potential for Biofertilizers

• Agronomic Benefits

The agronomic advantages of biofertilizers are substantial in the context of climate change, land degradation, and the decline in crop flavor quality. They can play an important role in preserving ecosystems, which are increasingly affected by soil and water pollution. For farmers, they can also offer economic benefits in the long term, by reducing the cost of maintaining soil fertility.

According to literature reviewed by Yakhin et al. (2017) and Faessel et al. (2014), known mechanisms of action of biofertilizers include:

- Stimulation of seed germination, root growth, plant establishment and development, photosynthesis, nutrient uptake (especially nitrogen and phosphorus), resistance to biotic stress, soil nitrogen and phosphorus metabolism, and microbial activity in soils;

- Reduction or mitigation of abiotic stress effects (drought, heat, cold, salinity, oxidation, mechanical or chemical stress) (Desfontaines et al., 2018).

For inoculated seeds, Plant Growth-Promoting Rhizobacteria (PGPR) multiply using seed exudates, colonize the emerging root, and eventually the entire root system. Mycorrhizae are symbiotic associations between plants and specific soil microorganisms (Fruchart, 2016, citing Bhattacharyya & Jha, 2011). There are also several links between agroecology and biofertilizer use, as shown in the following table.

Table 5: Connection between the 3 Agroecology Principles and Key Biofertilizer Features (Faessel et al., 2014)

Agroecology Principles	Biofertilizers and Biostimulants
Increase in biodiversity	Qualitative modification of microbial communities, increase in soil microbial activity
Strengthening of biological interactions	Enhanced plant-microorganism interactions, regulation of multiple plant physiological mechanisms
Closing of biogeochemical cycles	Improved nutrient absorption by plants, increased nutrient availability, stimulation of organic matter degradation

Source: Desfontaines et al., 2018

• Global Market Potential for Biofertilizers

The growing preference for organic food and increasing awareness of the risks posed by chemical fertilizers and air pollution have driven strong regional consumption, particularly in Europe. Strict regulations have also pushed many farmers to adopt biofertilizers over chemical alternatives (OCCITANIE TRIBUNE, 2019).

Globally, the biofertilizer/biostimulant market spans all five continents. Though representing just 0.6% of the global crop input market (Fruchart, 2016), it is growing rapidly, with annual sales growth of 12.5%, reaching over $2.2 billion USD in 2018 (Lambert, 2014). Europe leads the market, followed by North America. Combined with Asia-Pacific and South America, these regions account for over 50% of global revenue. "In Europe, biostimulants are mainly sold by SMEs. The market is estimated at €500 million, with use on approximately 3 million hectares" (Fruchart, 2016).

The market is segmented by product type: nitrogen-fixing, phosphate-solubilizing, etc. Nitrogen-fixing biofertilizers represented about 76.9% of global revenue in 2018, especially used for legume and non-legume crops, including sugarcane and rice (OCCITANIE TRIBUNE, 2019).

Historically, developing countries such as Cuba and India were pioneers in biofertilization. In Cuba, mycorrhizal inoculation became widespread after the collapse of the Soviet Union — its main fertilizer supplier (Tanguey, 2014, citing Febles et al., 2011). In India, the push was driven by rising food demand, tropical richness, and soil infertility (Tanguey, 2014).

Biofertilizer commercialization began mainly in North America and Europe, where environmental concerns and sustainable agricultural policies fostered adoption. Key global players include Lallemand Inc., Camson Biotechnologies Ltd., Novozymes, Symborg SL, National Fertilizers Limited, Agronutrition, among others. Market forecasts predict continued growth, driven by rising demand for environmentally friendly agriculture, reduced use of chemical inputs, and supportive regulations. According to TB Économie (2018):

- Rising demand across the agrochemical industry and increasing adoption of plant-based growth enhancers will drive biofertilizer market expansion.
- Declining arable land due to overuse of synthetic fertilizers will favor biofertilizer penetration.
- Growing consumer preference for organic food and increasing health awareness are key growth drivers.
- Strict EU regulations on toxic growth agents and 30% subsidies to farmers for sustainable agriculture will boost demand.
- Ongoing R&D efforts to improve plant productivity will also support industry growth.

6.1.2. Technical Components of a Biofertilizer

To produce a biofertilizer, it is first necessary to have the essential basic components, namely raw materials, microorganisms, and a method.

Raw Materials

They can be of animal or plant origin. They may also be of various types, including seeds, leaves, roots, and exudates from higher plants, for their contribution of secondary metabolites (such as plants from the Moringaceae and

Fabaceae families) (Desfontaines et al., 2018, citing Yakhin et al., 2017). Various species of algae (such as marine sargassum), protein hydrolysates, amino acids of animal origin including waste and by-products, humic substances, chitin or chitosan derivatives from insects, extracts from food or industrial waste, composts and compost extracts, vermicomposts, aquaculture residues, and treatment effluents are also used (Op. cit.). On farms, one or more categories of the aforementioned products can be used, particularly manure (from pigs, chickens, cows, horses, rabbits, etc.), discarded fruits and vegetables, green waste, and various plants known for their insecticidal, bactericidal, fungicidal properties, etc. (Desfontaines et al., 2018, citing Boulogne et al., 2012).

Microorganisms

They may be referred to as "effective microorganisms" (Desfontaines et al., 2018). These "mainly refer to mixed aerobic and anaerobic microorganisms, including various microbes such as lactic acid bacteria (Lactobacillus casei), photosynthetic bacteria (Rhodopseudomonas palustris), yeasts (Saccharomyces cerevisiae), and others" (Op. cit.). These microorganisms are generally selected from the local environment to ensure optimal effectiveness of the biostimulants, as they are adapted to the soil and climate conditions of the treated area (Groupe Eléphant Vert, 2018).

Methods

Methods vary widely depending on whether the producer is an individual or a company. Preparation and extraction technologies also vary—some are simple, others are very advanced, and some are specific depending on the nature of the raw material.

6.1.3. Selected Case Studies

The case studies examined are the biofertilizers developed by the Éléphant Vert group and Biotech Services Senegal.

• Biofertilizers by the Éléphant Vert Group

> To meet the needs of its clients, Éléphant Vert's expertise covers a wide range of industrial processes, including the production of soil amendments and organic fertilizers via composting platforms, the formulation of bio-inputs, and the production of micro-fungi and bacteria/yeast through liquid or solid-state fermentation. Éléphant

Vert's success is built on three key pillars: innovation, scientific partnerships & collaborative projects, and environmental respect (Éléphant Vert website, 2022).

The company offers a wide range of biofertilizers and biopesticides. Two organic fertilizers are presented here: FERTINOVA and ORGANOVA.

FERTINOVA (4-3-3) is an NPK fertilizer obtained by drying and sanitizing poultry droppings. Sold in 25 kg bags, it provides a balanced nutrient supply according to product descriptions. The nitrogen in the droppings is largely mineralized during the crop cycle, thus supporting crop growth while improving soil fertility, particularly by stimulating biological activity. A liquid version, FERTINOVA Liquid 3-3-8, is also available, derived from the valorization of agro-industrial co-products. It is designed for fertigation (the technique of supplying fertilizers through irrigation water), optimizing the nutrition of irrigated crops. This product is also approved for organic farming, in accordance with current European regulations (Op. Cit.).

ORGANOVA is an organic soil amendment sold in bulk and in 50 kg bags, which can be manually spread. According to product descriptions, it is made from composted plant residues, livestock effluents, and agro-industrial co-products. Its highlighted benefits include improved soil-plant interactions thanks to stable organic matter, maintenance of soil crumb structure, increased water retention, improved storage and availability of nutrients, and enhanced soil biological activity (Op. cit.). Like FERTINOVA, it complies with organic agriculture standards as per European regulations.

Background: Éléphant Vert is an international commercial group of Swiss origin, created by Antenna Technologies. Co-founder Sébastien Couasnet launched the group in 2012 by establishing the first industrial hub in Meknès, Morocco. Today, the company operates in Europe (France and Switzerland) and Africa (Morocco, Mali, Senegal, Kenya, and Côte d'Ivoire). According to its website, its goal is to promote large-scale, high-performance, healthy, and sustainable agriculture. In Senegal, the company was established in April 2017, located in the Almadies district on the road to Ngor Village in Dakar (Éléphant Vert website, 2022).

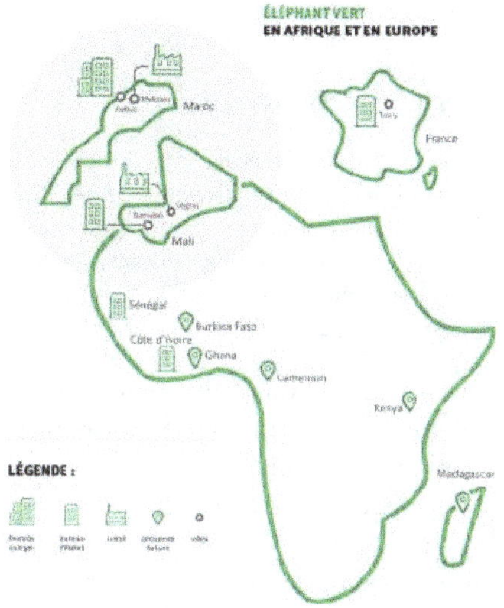

Fig. 13 & 14: Éléphant Vert's locations and products
Sources: commodafrica.com and elephantvert.com

To develop its biostimulants, Éléphant Vert has created a strain library (a certified collection of microorganism strains). Microorganisms are collected in areas with a high presence of specific pathogens. The trophic relationships between the pathogen and its neighbors (indifferent, additive, or antagonistic) are then studied. This results in a living collection of organisms with biopesticide or biostimulant potential. After conducting studies on "germ/activity correlation" and agronomic trials, the products undergo safety and regulatory approval before entering the market as agronomically valuable solutions. (Alasri, 2018, Managing Director of ÉLÉPHANT VERT Morocco)

• Biofertilizers from Biotech Services Senegal

> Founded in 2013, Biotech Services Senegal is an independent company under Senegalese law, [...that] provides products, services, engineering missions, studies, training, consulting, project management support, and agricultural production services in the fields of agro-environmental development and renewable energy (rural electrification, solar, biogas) within the framework of sustainable development (Biotech Services Senegal website, 2022).

The head office is located in Dakar at Sicap Sacré-Cœur 1, and the production plant is based in Thiès, at Keur Guilaye, near the monastery. Biotech Services Senegal also has distribution points in Mali, Guinea-Bissau, and Burkina Faso. According to its website, the products (biofertilizers, biopesticides, biostimulants) are locally made from 100% natural organic materials, in compliance with laboratory standards. These materials are selected and enriched to improve their absorption by soils and plants. This study focuses on the Bf2 biofertilizer, available in two forms: solid and liquid.

The solid Bf2 biofertilizer is produced from a blend of organic waste of animal or plant origin, selected for their agronomic value (especially their nitrogen, phosphorus, potassium, and micronutrient content). A bacterial ferment is added to the mix, which then undergoes pasteurization to eliminate pathogenic germs. The pasteurized product is composted with regular turning, then left to mature in windrows for 4 to 6 weeks. After maturation, the product is processed through an automated production line involving grinding, sieving, filtering, cooling, addition of natural pesticide, granulation, bagging, and weighing into 50 kg bags. This process is the result of a public-private partnership between Biotech Services Senegal, ISRA, Biopass (UCAD), and the French partner Biopost Cofuna (DIENG, CEO of Biotech Services Senegal, 2020). The 50 kg bag is sold at 10,000 FCFA. The recommended application is by spreading or digging, at a rate of 1 to 2 tons per hectare, depending on soil type.

REFERENCE ECHANT	15254 Engrais Vert
Constantes Physiques	
Humidité Résiduelle %	25.87
Cond µ/cm	6143.84
Calcaire total %	0.281
P_2O_5 Total %	0.147
Carbone Organique %	13.82
N%	1.66
PH (EAU)	8.1
PH (KCL)	7.9

Fig. 15. Overview and reference of solid Bf2 biofertilizer
Source: Biotech Services Senegal website, 2022

The liquid version of Bf2, named Bf2 HYFER Plus, is a foliar fertilizer composed of macronutrients that fertilize "all types of crops" (cereals, vegetables, fruit trees, flowers). Bf2 HYFER Plus is claimed to enhance plant resistance to diseases, increase production by at least 30%, and have no negative environmental impact. This liquid biofertilizer comes in two variants: Bf2 HYFER GREEN and Bf2 HYFER RED.

6.1.4. Identified Impacts

Overall, starting with Éléphant Vert, it can be noted that its production capacity (across all countries) is estimated at 100,000 tons per year of biofertilizers and 120 tons per year of biopesticides (Éléphant Vert website, 2022). In Senegal, in 2018, the group entered into an agreement with PDIDAS for the distribution of organic inputs in the northern zone, covering the municipalities of Gandon, Fass Ngom, Ndiébène Gandiol, Diama, Ngnith, Keur Momar Sarr, Syer, Mbane, and Ronkh. The agreement involved "a total of 125 tons of organic soil amendment for an experimental area of 42 hectares" (Ibid.).

On the company's website, there is also a testimonial from a farmer (Madiakhaté Niang) who stated that beyond the improved yields, his potatoes were more resilient and that "his fellow farmers were so satisfied" that they were determined to abandon chemical products.

Photo 56: 50 kg sack of Éléphant Vert

Source: Éléphant Vert website, accessed in August 2022

As for Biotech Services Senegal, its solid BF2 fertilizer is patented by OAPI under reference 120 19 00 dated May 27, 2019. This patent is valid in ECOWAS member states (Biotech Services Senegal website, 2022). The company has also received several international awards, including:

- Winner at the Forum of 100 Innovations for Sustainable Development organized by AFD on December 5, 2013 at the Elysée Summit in Paris;
- Selected by FoodByte and Rabobank among the top 10 projects worldwide out of 150 entries. The innovation was presented on October 25, 2016 at the University of Denver, Colorado (USA) and was later ranked 3rd;
- First place at the "Africa Climate Change Project Investor Forum" on February 16, 2017, in Nairobi, competing against 300 African projects. Following this forum, Biotech Services Senegal was selected to represent Africa at the Clean Energy and Climate Investment Forum on May 10,

2017, in Vienna, where it won the Best Project Award, ahead of five other competitors from various continents (Ibid.).

- Additionally, 700 tons of BF2 fertilizer were subsidized by the Government of Senegal for the 2021–2022 agricultural season. This quota was increased to 1,000 tons for the 2022–2023 season.

Photos 57 and 58: Production and distribution (Biotech Services Senegal)

Source: Biotech Services Senegal website, accessed in August 2022

6.2. Urban Micro-Gardening in Dakar

> Micro-gardens have proven to be well-suited for improving the availability of and access to quality vegetables for disadvantaged populations, thereby contributing to the fight against food and nutritional insecurity, which is on the rise in urban centers. UN-Habitat notes that global population growth will lead to an increase in the number of poor people living in slums and deprived neighborhoods on the outskirts of cities (FAO, 2010).

As the name suggests, urban micro-gardening involves growing vegetables in small spaces within the city. These spaces can include terraces, rooftops, balconies, courtyards, public squares, or other similar areas. It falls under the category of soilless agriculture and can be implemented using various reclaimed materials such as tires, barrels, crates, bathtubs, wooden pallet containers, etc. According to CEDD, *"the primary objectives of micro-gardening are to ensure a varied, diversified, and balanced diet nearby, and also to serve as a source of income for women working from home and for unskilled youth who struggle to find employment."* According to the FAO (2010), micro-gardens are particularly suitable for cities and especially for densely populated neighborhoods with no available land around homes. They allow low-income families to grow vegetables and various seasonings at home, enriching meals with essential vitamins and trace elements. The technology is simple, low-cost, and accessible, requiring little physical effort. Furthermore, micro-gardens operate in closed-loop systems, helping to conserve irrigation water and fertilizer. Soil fertility can be enhanced with compost made from household waste (FAO, 2010).

Among other advantages, the website amenagement-de-jardin.ooreka.fr highlights the high yields per square meter that micro-gardens can offer (up to 10 cabbages every 3 months, 36 lettuces every 2 months, 100 onions every 4 months).

Given these characteristics, it is reasonable to consider urban micro-gardening as a major innovation in the context of current climate change and urbanization—particularly for a "Global South" country like Senegal.

6.2.1. Study Area and Project Presentation

- **The Study Area**

The city of Dakar is the capital of the Republic of Senegal and the Dakar region. It is one of the country's four historic municipalities and was the former capital of French West Africa (AOF) (Wikipedia, 2022). Covering an area of 82.32 km² (0.042% of the national territory and 15% of the regional territory), the city had 1,363,444 inhabitants in 2019—representing 8.4% of the national population and 36.5% of the regional population. Its population density is 16,563 inhabitants/km² (Ville de Dakar, 2022). It is made up of the following 19 municipalities: Fann-Point E – Amitié; Yoff; Grand Yoff; HLM; Biscuiterie; Parcelles Assainies; Ouakam; Fass-Colobane-Gueule-Tapée; Grand Dakar; Hann Bel-Air; Dakar Plateau; Sicap Liberté; Ngor; Gorée; Patte-d'Oie; Cambérène; Médina; Mermoz – Sacré-Cœur; Dieuppeul Derklé Castors (Ville de Dakar, 2022).

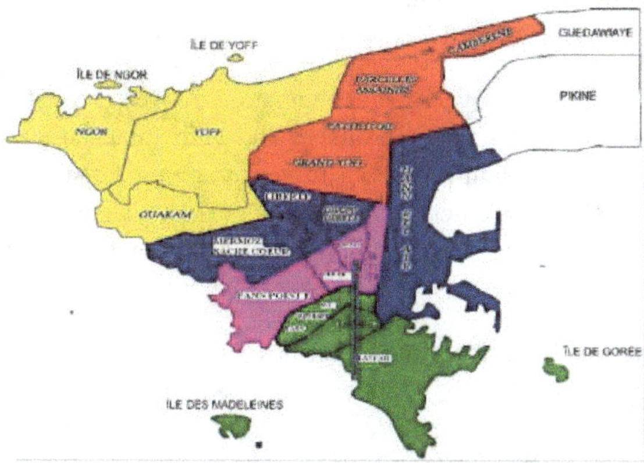

Fig. 16 & 17: Urban region maps of Dakar and the city's municipalities
Source: https://journals.openedition.org/cybergeo/27845 and City of Dakar (received in March 2022)

On this relatively small area, the city of Dakar concentrates important administrative, economic, political, and cultural functions that place it at the forefront nationally and also in the West African sub-region. The city's attractiveness has led to rapid urbanization, resulting in spatial expansion and a growing demand for urban services. This rapid growth, amplified by rural exodus and Dakar's role as the economic heart of the country (representing 80% of activities), has made it difficult to manage urban sprawl (Pouye, 2017). The loss of forest and agricultural spaces in favor of housing has had a negative impact on quality of life. Since the municipality's priority mission is to improve living conditions, it must address the challenge of feeding a growing urban population in a context where arable land is becoming increasingly scarce (Op. Cit.).

- **Dakar's Urban Micro-Gardening Project**

The history of micro-gardens in Senegal [...] is part of the FAO's "Growing Greener Cities" initiative. The micro-garden technology was introduced in the country—a first for Africa—through South-South cooperation. It was already known in Latin America. With support from the FAO's Technical Cooperation Programme (TCP/SEN/8823), the first training and demonstration center for micro-garden technology was established in Dakar in 1999, at the Center for Horticultural Development (CDH). Later, in 2003, the technology was integrated into the Special Programme for Food Security (SPFS) (SPFP/SEN/4501), and expanded to other regional capitals of Senegal under the National Food Security Program (PNSA) (FAO, 2010).

As part of the friendly relations and cooperation between the city of Milan and the city of Dakar (which are twin cities), and with financial support from the Italian Ministry of Foreign Affairs, a tripartite agreement was signed in May 2004 between the two cities and the FAO. This enabled the launch of a decentralized cooperation project to strengthen micro-gardens in the urban and peri-urban areas of Dakar.

The main goal of this project was to achieve social and food stability. It also aimed to improve nutrition by facilitating access to and consumption of a wide range of vegetables, while also serving as a source of income for underprivileged populations (Diop, 2016).

> The first phase of the project began in 2006 and achieved very significant results over about three years with a budget of US$465,000. The project received an international award for best practices in the horticulture sector, granted by the municipality of Dubai (Dubai Award), with a prize of US$30,000. This award helped reinforce the budget for a second phase and expand the model to other sub-Saharan cities (Ibid.).

(The first logo from the left represents Italian cooperation, the second and third represent the cities of Milan and Dakar, respectively. The last three are the FAO, NGO ACRA, and the University of Milan. Other stakeholders include public institutions such as the Ministry of Agriculture.)

After being informed and sensitized about the project by the city of Dakar, citizens take the initiative to apply for participation. The city has established training centers where they receive technical training and support (Mairie de Dakar, 2022). To ensure the sustainability of the activity, the City of Dakar has implemented several measures:

- Creation of twelve training centers throughout the city;
- Selection and training of instructors;
- Support for the micro-gardeners' association;
- Securing of gardening sites;
- Provision of equipment and water supply (not for all sites);
- Free training to promote wider adoption;
- Ongoing technical supervision and monitoring of the sites (Op. Cit.).

6.2.2. Technical Components of a Micro-Garden

A micro-garden requires a few essential elements: a support structure such as a tray, table, tire, bathtub, or any other type of recycled material; a growing substrate; water; seeds or young plants; and appropriate fertilizers. There are several cultivation methods used in micro-gardening. One common approach is growing on a solid substrate, typically in containers or tables, using a mix composed of 40% peanut shells, 40% rice husks, and 20% laterite, which together provide a lightweight and fertile base for plant growth. Another technique involves a floating system, where vegetables grow on sheets of polystyrene wrapped in foam, allowing them to float and absorb nutrients directly from the water below. This method is particularly efficient in conserving space and water. A third approach is soil-based cultivation, where traditional soil is enriched with compost or animal manure to improve its fertility and structure, offering a more familiar environment for plant roots. Each of these methods has its own advantages, and the choice often depends on the available space, materials, and the specific needs of the community or household using the micro-garden. The following photos provide a view of the micro-garden sites visited in Patte-d'Oie and Grand Dakar.

Photos 59 to 66: Illustrations (Patte-d'Oie and Grand Dakar centers)

Patte-d'Oie Center

Grand Dakar Center

Source: Author, 2022

6.2.3. Project Impacts

According to the Director of Planning and Sustainable Development for the City of Dakar, as cited in Debru et al. (2017), the achievements of the project include the fact that this food production system has become a model reference, and decision-makers have committed to its long-term sustainability by mainstreaming the practice and integrating the micro-gardening program into the city's development plan.

Since the technology is very basic and inexpensive, it is feasible to replicate the project throughout the entire city and Dakar region. It has also been expanded to other parts of Senegal thanks to state intervention and to other African countries through South-South cooperation. In this context, the cities of Banjul (The Gambia), Niamey (Niger), Praia (Cape Verde), Ouagadougou (Burkina Faso), and Douala (Cameroon) have benefited from training provided in Dakar for their technicians, as well as technical missions from Dakar to support the implementation phases (Interview Guide, 2022).

Moreover, it is noted that after just over ten years of existence, the Dakar micro-gardens have achieved the following:

- 12 training and demonstration centers (CFD) established across the city;
- Around 10,000 people trained for free through a 5-day course, followed by technical support;
- More than 10,000 planting containers distributed free of charge to beneficiaries;
- 147 community production centers (CPCs) created, allowing beneficiaries to gather on designated sites—generally made available by the municipality—across the 19 municipal districts of Dakar;
- Strong gender inclusion, with 83% of beneficiaries being women, and 50% under the age of 36;
- Training of vulnerable groups, including people with disabilities (1 dedicated center), prisoners (1 prison facility), and the elderly;
- Engagement of 25 elementary schools, with training provided to teachers, cleaning staff, security personnel, and students;
- Establishment of 4 sales kiosks made available to beneficiaries (though not currently operational) (Interview Guide, 2022).

To recap, the project unfolded in three main phases:

- Phase 1 (2006–2008): focused on consolidating the foundations of micro-gardening in the city of Dakar;
- Phase 2 (2010–2013): addressed the issue of product commercialization;
- Phase 3 (2013–2015): aimed to improve access to inputs, including the establishment of the Cambérène central purchasing center and the expansion of the project across the West African sub-region (City of Dakar, 2022)

It is important to note, based on field surveys and interviews conducted in 2022, that this data is no longer up to date. The mentioned CFD centers are not all operational, nor is the procurement center in Cambérène. No updated figures regarding active beneficiaries are available. Moreover, the previously reported impact data is also under discussion — a point that will be explored further in the final part of the results analysis.

Conclusion of Chapter 6

This chapter provided an analysis of two major innovations in sustainable agriculture: the biofertilizers developed by Éléphant Vert and Biotech Service Senegal; as well as urban micro-gardening in Dakar. The study highlighted their agronomic relevance, technical feasibility, dissemination potential, and impacts within specific contexts. These findings lay the groundwork for considering their integration on a larger scale.

Conclusion of Part Two

The objective of this second part was to present the innovations studied in each sector, along with their areas of dissemination and/or the stakeholders involved. For each innovation, a technical description was provided, accompanied by the recorded impacts. Only the positive impacts, based on data provided by the stakeholders, were presented. Negative impacts, encountered difficulties, and challenges will be addressed in the third part, which is dedicated to the analysis and interpretation of the results.

Part Three

Analysis and Interpretation of the Results

Introduction to Part Three

After presenting the global and national Senegalese context related to the three selected sectors, along with the ecological innovations studied, this third part returns to the initial question: Why do seemingly successful and positively assessed actions struggle to scale up in so-called "underdeveloped countries"? To answer this, hypotheses were proposed, serving as a guiding thread for this section's development.

This part now focuses on analyzing (explaining) and interpreting (uncovering what is not explicitly stated) the results. Chapter 7 will attempt to test the hypotheses, while Chapter 8 will present a broader discussion of the study by addressing ambiguities identified and situating the findings in relation to existing research in the field.

Chapter 7

Testing the Hypotheses

Partial Introduction to Chapter 7

Chapter 7 offers a critical analysis of the results, based on two main hypotheses.

- The first hypothesis suggests that certain ecological innovations, while accessible and beneficial to communities, struggle to spread widely, which limits their adoption and sustainability.
- The second hypothesis highlights structural barriers, particularly a mismatch between the content of certain innovations and the realities of the territories where they are implemented, thus hindering their large-scale integration.

7.1. Hypothesis 1: Some ecological innovations may be accessible and improve living conditions in multiple areas, but they suffer from diffusion issues, resulting in limited geographical adoption and lack of sustainability in the few areas where they are applied.

Hypothesis 1 can be considered validated if we focus on the word "some," which introduces a limitation and avoids overgeneralization. However, if we consider the broader formulation, particularly the diffusion challenges leading to weak adoption and lack of sustainability, the hypothesis becomes more debatable or even inaccurate in certain cases (to be clarified in Chapter 8).

As for the assertion that "some" innovations are accessible and bring positive outcomes yet struggle with diffusion, among the six innovations analyzed, two align with this pattern: Nubian vault construction and biofertilizers.

The Nubian Vault

Using mainly earth as raw material, the Nubian vault presents a potential alternative to concrete. However, this technique remains underused nationally. The project is currently concentrated in northern Senegal (Podor, Matam, Kanel), and its diffusion relies on development partners such as AVN, the Yvelines Departmental Council, and the NGO Le Partenariat. While local public authorities are increasingly supporting this technique (notably through public building contracts), several challenges persist. According to Le Partenariat (2021), obstacles include:

- Regulatory and insurance issues with earthen construction;
- Lack of qualified labor.

Document analysis and field visits revealed few individual homes built under this project. In an interview with a specialized mason in Podor, it was noted that some private orders exist, typically from wealthier individuals, but they are rare. Most constructions are initiated by local authorities, associations, or NGOs, whether local, diaspora-based, or international. From this 2020 summary, we observe that out of 22 buildings, classrooms, health infrastructure, and offices/housing dominate, each with four constructions. They are followed by departmental offices (3 buildings). Yet, one key goal of AVN's large-scale dissemination program "For Earthen Roofs in the Sahel" is to foster the development of a local sustainable housing sector in West Africa, offering affordable and adapted buildings, while promoting green job creation (Op. Cit.). This goal was not fully met in the VN Fouta project, particularly in terms of access for individuals. Still, efforts made by promoters deserve recognition. The four-year project (2017–2020) aimed to promote a variety of stakeholders (masons, businesses, architects, technicians, consultancies) trained in the Nubian vault technique. While it's unclear whether all objectives were met, significant progress was made in training: 21 level C4 masons (able to manage a site independently) and 172 apprentices trained between 2017 and 2020.

The Question Remains: Why Are Individual Constructions Still Rare?

A beneficiary-focused questionnaire was planned but received no responses. Hypotheses point to:

- Workload and costs as major obstacles;
- Despite lower structural costs, large amounts of earth and water are needed;
- Urban or resource-scarce areas pose logistical problems;
- Labor costs remain a factor, as with conventional building;

Though ecological advantages and longer durability are recognized, adoption also requires financial means and alignment with a specific ecological mindset. Stronger support might encourage adoption. For example, in Podor, soil used on-site was extracted mechanically and trucked in. If similar assistance were extended to private citizens, broader adoption could follow. Overall, the diffusion appears more institutionally focused than private. The State has not played an active role in popularization.

Biofertilizers

A slightly different dissemination dynamic emerged. The State did contribute:

- In 2018, 125 tons of biofertilizer were distributed to Éléphant Vert under the PDIDAS program;
- 700 tons of BF2 to Biotech Services Senegal for the 2021–2022 season;
- 1,000 tons for 2022–2023.

However, major challenges remain. These include a lack of visibility and information about biofertilizers, particularly in rural areas, and the absence of responses from Éléphant Vert despite outreach efforts. Although Biotech Services Senegal's Technical Director identified the northern and southern regions of Senegal as potential study zones, no mapping was completed due to time constraints. On the production side, collecting sufficient raw materials remains difficult, and financial constraints further hinder adoption. A 50 kg bag of biofertilizer is sold at 10,000 FCFA, which, while seemingly affordable, is prohibitively expensive for most small-scale rural farmers—especially considering that one bag is insufficient to cover an entire field. For instance, the recommended dosage per hectare is 1 to 3 tons for FERTINOVA (Éléphant Vert) and up to 6 tons per hectare for ORGANOVA. At this rate, costs can reach up to 600,000 FCFA per hectare, a significant burden during the lean season when daily expenses are already hard to meet.

In contrast, more accessible alternatives exist in the form of self-produced biofertilizers, such as those developed by the A.P.E. Caraïbe association or the locally adapted solution observed in Ndianda, in the Mbour department. These methods typically rely on freely available or locally sourced materials, making them better suited to the Senegalese context. Similar techniques were also discussed in Chapter 3 of Part I. Furthermore, local initiatives such as the Grand Dakar urban micro-gardening center have implemented organic fertilization methods, and in Toubacouta, the Nebeday association has integrated an agroecological component into its programs, though detailed information was limited due to a primary focus on biochar. Despite the promise of these approaches, their effectiveness at a larger scale still needs to be properly evaluated. In this regard, Biotech Services Senegal's Technical Director has proposed the development of a mixed (organo-mineral) fertilizer to help bridge the gap between accessibility, affordability, and agronomic efficiency.

7.2. Hypothesis 2: Other ecological innovations, by contrast, contain elements (technical, financial, social, or otherwise) that are not compatible with the realities of the host territories, thus limiting their widespread adoption

This hypothesis can also be considered as validated, although certain nuances are further developed in Chapter 8. The use of Compressed Earth Blocks (CEB) and biochar provides relevant illustrations of this dynamic.

Although CEB construction may initially appear straightforward, it actually requires substantial resources. Two main options are available to those wishing to adopt this technique: producing the blocks themselves or purchasing them from a third party or specialized company. In the first case, it is necessary to invest in an appropriate press and acquire basic technical know-how. In the second, purchasing the blocks requires significant financial means. In both cases, purchasing power remains a central issue, alongside the technical skills required. This explains why buildings constructed with CEB are mostly commissioned by private individuals with significant capital, private organizations, or public institutions.

In a 2021 report broadcast by TV5 Monde, Doudou DEME, founder of Elementerre, provided insight into the difficulties of disseminating this technique. He noted that:

"The average Senegalese wants to build as quickly and cheaply as possible. So when they decide to build, they'll go with the most familiar solution—and what they always see is concrete [...] Public authorities prioritize cost over quality."

This comment highlights the low visibility of alternatives to conventional concrete and the institutional inertia that hampers ecological innovations. While this analysis is not entirely inaccurate, it must be complemented by acknowledging the crucial role of cost, as DEME himself noted.

During an interview with a representative of Village Pilote Senegal (Mr. LOÏC), a question was raised about comparing the cost of a traditional concrete building and one made with CEB. The response indicated that costs are generally similar, as CEB is often only used for infill, while the foundations and reinforcements are still made with reinforced concrete. The lack of qualified labor for this type of construction was also emphasized. This hybrid approach (concrete + CEB) was observed at the Vocational Training Center in Gossas (Fatick region), and several examples in Chapter 1 also reflect this reality.

Photos 67 and 68: Vocational Training Center in Gossas built with CEB

Source: Author, June 2022

It is noteworthy that the structure (in grey) is made of reinforced concrete, while the infill (reddish-brown) uses compressed earth blocks.

Additionally, there was no use of raw earth in the CEB production process; most actors primarily used laterite stabilized with cement. This becomes problematic in areas where laterite is not locally available. According to Mr. LOÏC of Village Pilote, sourcing laterite requires supply from various regions. As such, CEB as an ecological innovation continues to face logistical and structural obstacles.

A similar situation arises with biochar. While the concept of an ecological charcoal alternative may seem appealing, both technical and cultural challenges persist. Traditional wood charcoal remains the dominant domestic fuel in Senegal, favored for its availability and widespread use. Any alternative must match or surpass it in performance while remaining affordable.

However, survey results indicate that biochar has a lower calorific value than wood charcoal. Users reported that it is fragile, hard to ignite, unusable when wet, and produces large quantities of ash. Without proper demonstrations or support, these limitations can lead to a negative perception of the product. In a 2020 MEDI1TV report, a woman expressed skepticism about using biochar for ironing, stating that a friend had found the excessive ash ruined her clothes. Similarly, the Nebeday association noted that the main obstacles are more cultural than technical. They stated: *"It remains difficult to change people's consumption habits. There is a need for education and support. Often, households buy straw charcoal to try it out and later say it doesn't work because they don't know how to use it properly."*

Therefore, awareness-raising, information sharing, and user support are essential steps for promoting the adoption of this alternative fuel. As recommended in a 2022 fact sheet by the association, clear guidance on how to use the charcoal should be provided before sale—through training sessions or participatory cooking workshops.

While some elements of this position are valid, they also expose the technical limitations of biochar as an alternative energy source. In response to these constraints, Nebeday has redirected its efforts toward a specialized version of biochar—Kërin Naatangué Thiouraye—used specifically for incense and tea preparation. The original version, Kërin Naatangué Tiger, initially designed for cooking meals, has since been discontinued. This reflects a more segmented market approach.

Overall, the examples of CEB and biochar illustrate the challenges that hinder the adoption of certain ecological innovations. These challenges are technical, financial, and cultural in nature. While they have not led to the total abandonment of these solutions, their diffusion has remained limited to a narrow segment of the population—typically wealthier individuals or those particularly attuned to environmental issues. This group remains marginal compared to the original target audience, which included more vulnerable social groups.

Partial Conclusion of Chapter 7

This chapter has examined the two hypotheses formulated around the adoption of ecological innovations. Regarding the first hypothesis, it has been confirmed that some innovations, while useful and potentially accessible, suffer from poor dissemination strategies, which hampers their widespread adoption and compromises their sustainability—even in the areas where they are being tested. The case of Compressed Earth Blocks (CEB) clearly illustrates this trend, with adoption hindered by a lack of visibility, financial barriers, and insufficient awareness.

The second hypothesis has also been validated through examples such as biochar. Some innovations contain technical, financial, or cultural features that are not well aligned with local realities. This misalignment makes them difficult for most of the population—particularly the most vulnerable—to adopt. As a result, despite good ecological intentions, these innovations struggle to establish a sustainable presence in the targeted territories. These findings highlight the need for more intensive adaptation, mediation, and support efforts to make such innovations visible, understandable, and truly accessible.

Chapter 8

Discussion

Partial Introduction to Chapter 8

This chapter offers a critical reflection on the findings obtained through the analysis of the ecological innovations studied. The aim is to explore the ambiguities observed in other innovations not previously addressed in the verification of the hypotheses. Subsequently, a broader contextualization of the results is provided by comparing them with existing work in the field. This helps position the dissertation within the wider scientific landscape of innovation, sustainable development, and territorial ecology, while also identifying the specific contributions of this research.

8.1. Ambiguities in the Results

It is worth noting that the Taiba Ndiaye Wind Farm (PETN) and urban micro-gardening in Dakar were not discussed under either of the two hypotheses defined earlier. This represents one of the key ambiguities of the study.

In the case of the Taiba Ndiaye Wind Farm, it cannot be definitively asserted that the project is inherently incompatible with the realities of all potential host territories in Senegal, thus preventing its replication. The issue of the project's replicability in other areas of Senegal was addressed in the interview guide submitted to the company overseeing the PETN, with the question framed as: "Do you think this type of project could be easily replicated in other areas of Senegal?"

The response was:

> "Yes, subject to several parameters, notably the availability of wind at acceptable thresholds. For now, the potential of the 'Grande Côte' region from Dakar to Saint-Louis, swept by the Harmattan and Atlantic winds with speeds ranging from 8 to 20 meters per second (m/s), is ideal for wind farm installation. Senegal, with its 700 km of coastline, could also consider offshore wind turbines located at sea rather than on land, to make better use of wind energy and produce electricity via a turbine and an electric generator." (PETN Interview Guide, July 2022 – Author's translation)

Several key insights can be drawn from this response. First, the installation of a wind farm requires specific conditions, among which wind availability at viable speeds is a critical factor. This requirement renders areas lacking sufficient wind regimes unsuitable—indicating a form of territorial incompatibility. However, the mention of Senegal's 700 km of coastline suggests that other regions remain favorable for hosting wind energy projects.

By incorporating additional parameters—technical, technological, and financial—it becomes evident that this type of project primarily fits within an institutional or public-private partnership framework, where the state and multinational companies play central roles. The involvement of local populations remains limited, mostly confined to how they perceive (positively or negatively) the project. Therefore, when such an initiative is considered an ecological innovation, it can partially confirm both hypotheses, contradict them in certain respects, or even fall outside their scope in specific contexts.

In contrast, a different pattern emerges in the case of urban micro-gardening in Dakar. Unlike wind energy, this innovation can experience either high or low levels of adoption, as well as varying degrees of sustainability over time. The basic elements required to establish a micro-garden—containers, substrate, seeds or seedlings, and fertilizers—remain partially accessible to urban populations, although financial constraints may arise in certain contexts.

On this point, women interviewed at the Centre de Formation et de Développement (CFD) in Patte d'Oie reported difficulties related to water access: their well had dried up, and they needed to make financial contributions to refill the electric meter that powered the pump. This issue was not mentioned at the CFD in Grand Dakar, where the system is connected to the municipal power grid.

Generally speaking, the main obstacle reported by those surveyed was the lack of motivation among trained beneficiaries to continue the activity independently and sustainably. A significant number of participants abandon the practice shortly after the training, even when it is offered free of charge.

From the map, it is apparent that the number of active beneficiaries during that period did not exceed 500 individuals. When asked for updated figures in 2022, no response was provided. It is therefore assumed that among the 10,000 people trained, a large portion discontinued the activity.

Furthermore, at the CFD centers, the current vegetable output remains low. During field visits, the main crops observed were mint and lettuce. At Patte d'Oie, vegetables were reportedly abandoned due to rat infestations, and mint cultivation proved to be more financially profitable. It was also noted that the initiative did not attract much interest from youth; the participants encountered were mainly older women, with a few women in their thirties or older.

Thus, it can be concluded that other ecological innovations, even when accessible, may face adoption difficulties that are not due to the nature of the innovation itself, but rather to the characteristics of the target populations. These may include lack of interest, low motivation, time constraints, or other barriers that limit engagement. This observation could form the basis of a new paradigm in current research, challenging the assumption that "ecological" solutions are inherently desirable or widely embraced.

8.2. Positioning the Thesis in Relation to Other Research

It is important to emphasize that past and current research is so abundant that compiling an exhaustive inventory appears difficult. With the development of the internet, this trend has only intensified. Thus, the objective of this subsection is not to present a complete overview of existing work, but rather to offer an analytical reading based on the sources consulted in the field. We can say that our research aligns (fully or partially) with some works in the field of innovation diffusion (in general), and diverges from others. The focus will be placed more on those with which there is alignment.

Among the studies consistent with our results is Treillon (1992) on technological innovation in the Global South, using the example of the agri-food sector. The author's approach was to analyze several innovations (oil presses, grain mills, infant flours, cassava chips, palm sugar, etc.) while highlighting the limitations of traditional guarantee criteria for some of them. This led him to propose a different project management approach, positioning the diffusion of innovations as a lever for social transformation in support of more equitable development. Although the contexts differ, the statement that "for an innovation to be effective, it must combine and conform to the social, commercial, and technical realities of the environment in which it is introduced" resonates with our study. This was verified in the cases of biochar, compressed earth blocks (CEBs), and Nubian vaults.

However, this assertion cannot be generalized to all innovations. Some innovations—whether ecological or not—can profoundly disrupt preexisting social, commercial, and technical structures. Examples include technical advances like the wheel, railroads, automobiles, wired and wireless communication technologies, electricity, and aviation. These now-commonplace innovations have radically transformed societies. Therefore, other factors must also be considered to ensure the successful diffusion of innovative initiatives.

Continuing Treillon's line of thinking, partial convergence can also be noted with the work of Cécile, Warrant, and Valenduc (2001) on technological innovation for sustainable development. These authors identified several types of barriers that must be overcome for an innovation process to succeed: conceptual, organizational, knowledge-based, technological, and financial. This perspective is relevant to some of the ecological innovations observed. However, differences also emerge, particularly regarding the idea that innovation itself is not a source of change, but that its diffusion process determines its impact. According to these authors, this process rests on two fundamental dimensions: the intervention of change agents and the perceived characteristics of the innovation by adopters. The role of the change agent is defined as multifaceted: initiating contact, providing support, listening to needs, persuasion, and empathy.

Quoting Rogers, the authors emphasize the importance of certain qualities an innovation must have from the adopters' perspective: relative advantage, compatibility, perceived complexity, trialability, and observability. However, we argue that some innovations do not require a highly elaborated diffusion process because they do not directly involve individual adopters but operate more collectively—such as a wind farm. In this model, the innovation itself should be considered before its diffusion. A standalone diffusion without genuine innovation beforehand may not lead to change. Furthermore, as previously shown, there are innovations where the issue lies neither in the innovation nor its diffusion, but in the target audience. These individuals may simply not be interested in the product being offered.

To conclude this section on aligned research, we should mention Laval's (2014) study on green charcoal, examining its promises versus its realities as an attractive alternative. As an independent consultant, Laval evaluated a biochar project in Mali funded by the Agence des Micro-Projets (AMP-France), along with similar initiatives in Cambodia, Senegal, and Uganda, to capture lessons from failures, successes, and challenges. The Malian project was the primary focus in terms of methodology and detail. In 2011, AMP "took a gamble" by financing the Malian association Yasagu in a project to produce and commercialize charcoal briquettes from biodegradable waste.

The project aimed to create a Household Waste Recovery Center (CVDB) in the commune of Bandiagara to produce green charcoal and organic compost. The CVDB specialized in collecting used paper and cardboard, wood chips, and other dry cellulosic waste. The project had a dual objective: combating deforestation through green charcoal and compost production, and improving waste management in semi-urban areas. After AMP's funding ended, Yasagu continued to work on improving the model. However, the CVDB still faces significant challenges preventing it from reaching financial sustainability: the quality of the green charcoal produced is not sufficient to convince users, and waste collection remains very costly, heavily impacting production costs and the final price for households. (Laval, 2014)

Two conclusions are drawn in this study:

First, "in an innovative project like Yasagu's, it is essential to set modest, achievable goals," and that "key pitfalls include overestimating the available raw materials, oversizing production equipment, and underestimating market penetration difficulties."

This is particularly relevant today, when terms like environmental protection, sustainability, organic products, and ecology are often overly valorized. While recognizing the urgency of environmental challenges, it is also essential to keep a realistic and balanced approach. Resources like oil, concrete, and coal did not become dominant by accident: despite their drawbacks, they are highly effective in their domains. Their potential alternatives must therefore offer comparable or superior performance, unless users consciously accept certain shortcomings—which is not yet the case for everyone.

A relevant illustration comes from a solar mini-power station visited on October 17, 2022, in Keur Babou Ndity, a village in the Fatick region of Senegal. A photovoltaic plant was installed to serve about 980 people, replacing connection to the national electricity grid. According to the village chief, Amadou Diallo, project promoters had promised that standard household appliances (lights, fans, fridges, mills, etc.) could be used without issue. However, during the visit, it was noted that electricity was only available from 1 PM to 9 PM, with total outages outside this window. The existing mill still ran on diesel. Ironically, several residents, including the chief, expressed a desire for "real electricity," meaning conventional electricity generated from fossil fuels.

Similarly, the case of electric vehicles can be mentioned. While they offer hope for environmental progress, several challenges must be considered: affordability in developing countries, availability of key materials like lithium and copper, and the energy impact of their production and operation. The war in Ukraine has also revived debates around global energy security. Furthermore, the geopolitical implications of rare mineral extraction, tensions in mining regions, and related environmental consequences cannot be ignored. Research conducted in a Master's thesis on the domestic biogas program in Senegal also showed that so-called "green" projects often suffer from poor implementation—mismanagement, embezzlement, and fund capture. Hence, it is crucial to balance ecological ideals with rigorous technical, social, and economic evaluations to ensure the viability and credibility of proposed alternatives.

Second, Laval concludes that "each situation is unique and generalization should be avoided." This resonates with our findings: each territory has its own specific context. An action may be considered innovative in one setting but not in another due to a variety of factors. Innovations such as Nubian vaults, compressed earth blocks (BTC), urban micro-gardens, wind farms, biochar, or biofertilizers all serve as relevant examples.

Partial Conclusion of Chapter 8

The critical analysis conducted in this chapter has revisited the results obtained through the study of ecological innovations, while also addressing the limits, ambiguities, and gaps left by some initiatives not included in the hypothesis-testing phase. This reflective review has highlighted the complexity of innovation adoption, especially when such innovations are introduced in heterogeneous sociotechnical and cultural contexts.

By comparing the data collected with existing research, this study is positioned within a broader scientific framework at the intersection of innovation studies, sustainable development, and territorial ecology. This positioning has brought to light the specific contributions of the research, particularly in analyzing the discrepancies between the initial intentions of innovation promoters and the actual dynamics of reception by the targeted populations.

Without claiming to be exhaustive, the discussion in this chapter underscores the importance of adopting a critical, contextual, and interdisciplinary approach to understanding the trajectories of ecological innovations. This opens the door to a more nuanced reflection on the conditions for the success of these innovations, while challenging some common assumptions about their diffusion, effectiveness, and social legitimacy.

Conclusion of Part Three

This third part has made it possible to test the initial hypotheses, engage in a critical discussion of the results obtained, and compare them to a body of existing scientific literature. This entire process has enriched the understanding of the conditions for the deployment and diffusion of ecological innovations in the territories studied.

It becomes clear that it is not appropriate to rely on a binary interpretation, as initially suggested by the two hypotheses, to explain the limitations faced by certain so-called innovative ecological actions.

General Conclusion

The study conducted on ecological actions in the sectors of eco-construction, alternative energies, and sustainable agriculture in Senegal has shed light on a variety of issues and dynamics. Two hypotheses were formulated at the outset in order to better understand the challenges faced by these initiatives: one focused on the difficulties of innovation diffusion, the other on their lack of compatibility with the technical, economic, and social realities of the host territories.

The first part provided a general framework through a presentation of the Senegalese context and the geographical challenges related to the sectors under study. The second part focused on a detailed analysis of six ecological innovations, exploring their technical characteristics, the profiles of the actors involved, their implementation areas (for some), as well as the achievements and impacts recorded. Finally, the third part was devoted to a critical analysis of the results, their contextualization, and their comparison with prior scientific research.

This analysis reveals that the initial hypotheses are confirmed in some cases but not systematically. Some of the innovations studied fall partially or simultaneously under both hypotheses, while others fall outside these explanatory frameworks altogether. Each initiative has its own specificities, making generalization difficult, despite the recurrence of certain factors such as financial constraints.

Overall, the ecological innovations examined appear to be relevant within the Senegalese context. However, their level of maturity and deployment remains insufficient to hope for a widespread and effective replacement of current practices in the sectors concerned—with the possible exception of biofertilizers, if their production were better mastered and internalized by the beneficiaries themselves.

This does not mean these initiatives should be abandoned. A rational approach, grounded in a nuanced understanding of local realities and territorial dynamics, remains essential. These projects must be viewed as steps in a broader process of learning, experimentation, and continuous improvement. Such efforts, supported by complementary research, contribute to the advancement of knowledge and practice, as has been seen throughout the historical evolution of technical and technological innovations worldwide.

Moreover, it is neither necessary nor desirable to replace everything. Hybrid approaches, combining ecological innovations with existing knowledge, may constitute viable alternatives. In some cases, the goal should not be to replace, but to improve or "re-innovate" based on what already exists.

Study Limitations

The study presents several limitations, which are important to acknowledge in order to better understand the scope and significance of the findings.

First, the analysis focused on six (6) ecological innovations across three sectors, two per sector. This methodological choice, while allowing for a certain diversity, cannot claim to cover all the challenges related to ecological innovations in Senegal. It represents a sample intended to identify some of the challenges—but not an exhaustive or representative overview.

Second, some innovations, such as biofertilizers and the Taiba Ndiaye Wind Farm (PETN), were analyzed using secondary data (articles, videos, websites, institutional documents, and interview guides), due to the absence or limited availability of first-hand empirical data.

In the case of PETN, a partial site visit was conducted in May 2022, without an expert guide. While some turbines were operational and others idle, this simple observation did not allow for an evaluation of actual power output or any associated technical or social challenges. A guided tour, later proposed by the company, could not be arranged in time due to difficulties obtaining authorization and scheduling a third field visit. The Taiba Ndiaye town hall did not respond to interview requests, despite two visits on-site.

Regarding biofertilizers, the situation was similar. No physical sample could be observed, and no beneficiaries were directly interviewed. Attempts to contact the companies Éléphant Vert and Biotech Services Sénégal were unsuccessful under favorable conditions: the former did not respond to calls or emails, while the latter replied briefly without the possibility of a deeper follow-up. Lack of time prevented site visits, although this option had been considered during initial phone contact. Responses often led to a request for written questionnaires, which yielded no concrete follow-up.

Another key limitation was the low response rate to the interview guides. Of the eight (8) guides sent by email or WhatsApp, only three received responses: the City of Dakar (urban micro-gardening, after about one month), Biotech Services (after three months and simplification of the guide), and PETN (five months later). Other actors (Nebeday, Elementerre Dakar, AVN-Fouta, Village Pilote, Taiba Ndiaye town hall) did not respond, despite multiple follow-ups. When written responses were not obtained, in-person interviews were arranged when possible.

This low response rate led to several methodological observations. It was noted that many actors, particularly project initiators, had limited time to dedicate to academic research. A commonly heard phrase during initial contact was: "Send the questionnaire, we'll see later." Furthermore, the first versions of the interview guides—judged too long or unfocused—may have discouraged some respondents. Informal feedback suggested that broadly worded questions (e.g., "Can you tell us about your organization?") were perceived as time-consuming. After the guide was simplified and reduced to a single page, response rates improved.

Another notable constraint was the inability to access beneficiary lists for certain innovations (Nubian vaults, biochar, urban micro-gardening in Dakar, biofertilizers), despite repeated requests to the relevant stakeholders. In response to this gap, alternative strategies were used: analysis of audiovisual testimonials, comparisons with press articles, direct field observations, and both formal and informal interviews with experts and non-experts in the relevant fields.

These various limitations, while they restricted the scope of the investigation, also provided valuable insights into the practical conditions of field research on innovations. Other limitations not identified here will likely be pointed out by future readers, which would offer opportunities for complementary or follow-up research.

Perspectives

Numerous perspectives could be considered.

From a comparative standpoint, it could be useful to expand the study to include other ecological innovations or other Global South countries, in order to identify both common patterns and territorial specificities. On a smaller scale, a more focused approach targeting a single sector, such as eco-construction, would allow for the search for concrete, adapted, and economically viable solutions that could be adopted at a lower cost. This type of approach aligns with a Research and Development logic applied at the local level, similar to what is observed in entrepreneurial environments.

In this spirit, some personal experiments in eco-construction have been conducted, particularly around the use of raw earth (earth-based construction). There is still room for technical improvements, especially by adapting the structural designs. For instance, increasing wall thickness and using overhanging roofs could significantly improve resistance to humidity. In the field of earth construction, it is often said that the golden rule is to have "boots and a hat"—that is, solid foundations and good roofing. This principle was observed in the Nubian vault technique. The thickness of the walls, as seen in examples like the vaults of the Ramesseum in Egypt, appears to play a crucial role in durability. Unlike some local constructions made with banco (mud), where the walls are often as thin as those made of cement, these massive structures offer better resistance to harsh weather, particularly during the rainy season.

Still within this framework of contextual innovation, it would also be valuable to revalorize materials generally considered pollutants, such as plastic. Some inspiring initiatives already exist. For example, in Cameroon, the company BOCOM Recycling manufactures eco-friendly roofing tiles from a mix of plastic and sand. According to the promoters, these eco-tiles are not only thermally and acoustically insulating, but also offer a longer lifespan than traditional metal sheets. If these performances are confirmed and if the product is financially accessible to low-income populations, such an innovation would deserve wider dissemination across the continent.

Beyond roofing, plastic recycling could also be applied to the creation of other construction or furnishing components: doors, windows, furniture, fences, or even modular housing structures. Several innovations in this direction are emerging in various countries, with some even experimenting with converting plastic into fuel.

This same innovative spirit could be applied to other sectors studied. During the interviews, some interesting ideas were proposed, such as the use of organo-mineral fertilizers or the development of offshore wind farms along the Grande Côte between Dakar and Saint-Louis. These proposals deserve further exploration.

Other potential research avenues include the valorization of rainwater and fecal sludge:

- Regarding rainwater, the significant amount of water that flows during the rainy season remains a largely untapped resource. It would be relevant to consider the creation of artificial reservoirs or retention basins in arid, saline, or water-scarce areas, including urban environments. Further thought could also be given to recovering surface water that currently flows directly into the ocean.
- As for fecal sludge and wastewater, their potential as organic soil amendments is often overlooked. In a context of rapid urbanization, the quantities produced are substantial, opening up possibilities for large-scale valorization projects. Some wastewater treatment plants, such as the one in the Niayes area, are already developing practices such as phosphorus recovery and water distillation. These experiences could be replicated in other locations.

In short, Senegal can be likened to a puzzle: the resources and capacities do exist, but they are scattered. What remains is to assemble them strategically and coherently, to bring forth sustainable solution.

Bibliography

1. Agronutrition. (2016). Unlocking the potential of nature [Press dossier].

2. Anne, B. N. (2003). Contribution to improving youth training in micro-gardening [Monograph, certificate of aptitude for youth education inspectors]. 64 p.

3. Agence Nationale de la Statistique et de la Démographie (ANSD). (2024). Provisional report Chapter X: Agriculture, Fifth General Census of Population, Housing, Agriculture and Livestock (RGPHAE-5). 46 p.

4. ANSD. (2016). Economic and social situation of Senegal in 2016 / Construction and Public Works. 11 p.

5. ANSD. (2014). Final report of the General Census of Population, Housing, Agriculture and Livestock (RGPHAE). 29 p.

6. AVN. (2021). Typology of Nubian Vault buildings in North-East Senegal. 23 p.

7. AVN. (2016). Manual of the Nubian mason – Class 4 and 5 version. 84 p.

8. AVN. (2014). Basic technical rules – Technical concept of Nubian vault (VN). 15 p.

9. Baud, P., Boulanger, S., & Béraud, C. (2013). Dictionary of Geography. Paris: Hatier. 607 p.

10. Belhedi, A. (2010). Epistemology of geography: Deciphering space. Tunis: University of Tunis, Department of Geography. 257 p.

11. Biotech Services Senegal. (2022). Catalogue. 8 p.

12. Brunet, R., Ferras, R., & Thery, H. (1993). The words of geography: critical dictionary. Montpellier, France: Maison de la Géographie, RECLUS Public Interest Group. 518 p.

13. International Labour Office (ILO). (2013). Local materials and eco-architecture in Senegal: status and perspectives within a green transition. ILO. 59 p.

14.Cécile, P., Warrant, F., & Valenduc, G. (2001). Technological innovation in the service of sustainable development. Belgium: Travail-Université Foundation. 82 p.

15.CGAAER (France). (2017). Support for Senegal in the reflection on the creation of agricultural chambers and agricultural and rural development (Report No. 16101). 34 p.

16.CIAT, BFS/USAID. (2016). Climate-smart agriculture in Senegal: Country profiles of CSA for Africa. Washington, D.C.: International Center for Tropical Agriculture; BFS/USAID. 20 p.

17.CNCR. (2013). Healthy and sustainable family farming (agroecology) as a model for agriculture in Africa. 38 slides.

18.World Heritage Convention. (2012). World Heritage: Inventory of Earthen Architecture. United Nations. 288 p.

19.Craterre. (2014). National program to reduce GHG emissions through energy efficiency in the building sector: Production project of thermal insulation materials based on Typha (Technical report). Ministry of Environment and Sustainable Development, 64 p.

20.Cros, V. (n.d.). Towards sustainable agriculture [Course]. CFPP A Ariège-Comminges. 12 p.

21.Debru, J., Albert, S., Bricas, N., & Conaré, D. (Eds.). (2017). Urban food policies: Proceedings of the international meeting on experiences in Africa, Latin America, and Asia regarding markets, collective catering and urban/rural linkages. Montpellier: UNESCO Chair Alimentations du Monde. 210 p.

22.Desfontaines, L., Rotin, P., & Ozier-Lafontaine, H. (2018). Biostimulants: What do we know? Alternatives for Guyanese agriculture. Innovations Agronomiques. 64 p.

23.Diallo, M. M. (2015). Extractive activities and territorial dynamics in Senegal: comparative study between gold and phosphate (Doctoral thesis). Paris 1 Panthéon Sorbonne; UGB. 343 p.

24.Diarra, N. (n.d.). Innovative systems in Senegal: biocharcoal. 4 p.

25.Diop, A. D. (2018). Challenges of ecological innovations in Senegal: Case of domestic biomethanation in Notto Gouye Diama and Bango (Master's thesis). UGB. 74 p.

26.Diop, M. B. (2016). Impact analysis of the Dakar micro-gardens project. Presentation at the Socio-Cultural Center of Grand-Dakar. 27 slides.

27.Dubois, J. M. (2004). Guide to writing a master's thesis in geographical sciences or environmental sciences (R type) and doctoral thesis in remote sensing. University of Sherbrooke. 93 p.

28.Enda Energie. (2016). Project ACE WA – Accelerating the implementation of regional policies on clean energy in West Africa. 52 p.

29.Enda Pronat. (2017). Ecological and organic agriculture initiative (AEB) led by the National Federation of Organic Agriculture (FENAB). 126 p.

30.Enda Pronat. (2019). Feasibility study: multi-actor partnership for agroecological transition. 35 p.

31.Fall, S. M. (n.d.). Investigation techniques, documentary research course. University Gaston Berger. 6 p.

32.Fall, A. M. (2009). Impact of solar cookers PCSA on conservation of ecological and social balance: Case of Ngaye Méckhé commune in Senegal. 28 p.

33.FAO. (2010). Micro-gardening manual for Senegal. 41 p.

34.Fondation Energie pour le Monde. (n.d.). Green electricity for thirty-five thousand rural people in Senegal. 52 p.

35.Fruchart, S. (2016). Tests of the effectiveness of different biofertilizers on spring barley (Hordeum vulgare) and natural meadow. University of Angers. 72 p.

36.Génération Agroécologique. (2016). Multi-actor approaches for the development of agroecology in West Africa. Report of the November-December 2016 days. 25 p.

37.African Development Bank Group. (2015). OMVG Energy Project. 41 p.

38.Guignard, M. (2006). Eco Construction. Conference of the Pyrenean Craftsmen. 40 p.

39.Guillaud, H., et al. (1995). Compressed earth blocks, volume II: Design and construction manual. 151 p.

40.Hugill, B. (2013). Biochar – a biological house for soil microbes. Technical note, ECHO. 5 p.

41.Lambert, L. (n.d.). Biostimulants: Let's talk about it. Agriculture, Fisheries and Food Quebec. 60 slides.

42.Laval, M. (2014). Green charcoal: hopes and realities of a seductive alternative. La Guilde Européenne du Raid. 69 p.

43.Legrand, B., Diem, M., & Gueye, M. (2017). Eco-construction and energy efficiency in buildings: good architectural practice manual. 91 p.

44.Le Partenariat. (n.d.). Nubian vault program Fouta 2017-2020. 4 p.

45.Le Partenariat. (n.d.). Eco-construction and sustainable housing component. 4 p.

46.Le Partenariat. (2018). Nubian vault project Fouta: activity report 2017. 30 p.

47.Le Partenariat. (2021). DEFI-VN Project presentation 2022-2025. 25 slides

48.(Author unknown). (n.d.). Description of the ecological villa "ASSIE GAYE". 7 p.

49.Le Partenariat. (n.d.). Access to renewable energy program – Northern Senegal Saint-Louis zone. 4 p.

50.Ministry of Agriculture and Rural Equipment (MAER). (2014). PRACAS Agricultural component of the Emerging Senegal Plan. 116 p.

51.Mbow, M. (2017). Challenges of Senegalese agriculture in a climate change perspective (Master's thesis). University of Sherbrooke. 79 p.

52. Ndao, M. (2012). Environmental dynamics and management from 1970 to 2010 of Senegalese wetlands: land use study by remote sensing of Niayes (Doctoral thesis). University of Toulouse. 370 p.

53. Nebeday. (n.d.). Straw charcoal: biocharcoal sheet. 5 p.

54. Nebeday. (2022). Biocharcoal monitoring July 2017–June 2018 [Excel file].

55. Nebeday. (2022). Dust production monitoring 2020–2022 [Excel file].

56. Nebeday. (2022). Production monitoring sheet 2019–2022 [Excel file].

57. Ndiaye, A. S. (2011). Experiences of PERACOD: promotion of alternative domestic fuels. PERACOD / Ministry of Energy. 21 slides.

58. International Labour Office (ILO). (2013). Organic agriculture in Senegal: status and investment potential in a green transition. 50 p.

59. Peyrot, D., & Istin, V. (n.d.). Eco-construction: technical sheet. 23 p.

60. Plateforme-ecco. (n.d.). What is eco-construction? 2 p.

61. Republic of Senegal. (2019). Exploratory report on the green economy. 106 p.

62. Republic of Senegal. (2010). Law 2010-22 on biofuels orientation. 8 p.

63. Republic of Senegal. (2015). National Renewable Energy Action Plan (PANER) Senegal, period 2015-2020/2030. 91 p.

64. Rodale Institute & Green Senegal. (2000). Towards regenerative agriculture in Senegal. 155 p.

65. Sy, K. (2014). Industry and territorial development: insertion of agro-industrial companies in the delta and lower valley of the Senegal River (Doctoral thesis). Paris Ouest & UGB. 306 p.

66. Tanguey, L. J. (2014). Mycorrhizal inoculants for a more productive and less mineral fertilizer-dependent Quebec agriculture (Master's thesis). University of Sherbrooke. 99 p.

67. Treillon, R. (1992). Technological innovation in southern countries: the case of agro-food. ACTT, CTA, Karthala. 267 p.

68. Trisha, G. (1994). Environment today. Singapore: Longman Group Limited. 112 p.

69. Val d'Oise. (2012). Eco-construction guide in Val d'Oise. 31 p.

70. Veyvet, Y. (Ed.). (2005). Sustainable development: plural approaches. Paris: Initial, Hatier. 287 p.

71. City of Dakar. (2015). Micro-gardens in the municipality of Dakar. 9 p.

72. Various authors. (n.d.). 38 dictionaries and correspondence collections [Digital dictionary].

Online resources

1. https://youmatter.world/fr/innovation-technologique-resoudre-crise-ecologique/ (Accessed on 02/02/2020)

2. http://www.environnement.gouv.sn/informations-sur-l-environnement/statistiques-g%C3%A9n%C3%A9rales (Accessed on 23/02/2020)

3. https://www.asn.sn/spip.php?article48 (Accessed on 06/03/2020)

4. https://www.researchgate.net/figure/Domaines-climatiques-du-Senegal-Sagna-2008_fig1_331232160 (Accessed on 23/11/2021)

5. http://www.dgpre.gouv.sn/index.php/2018/09/04/le-suivi-des-eaux-de-surface/ (Accessed on 23/11/2021

6. https://www.fao.org/3/a0970f/a0970f12.htm (Accessed on 23/11/2021)

7. https://fr.wikipedia.org/wiki/Dakar (Accessed on 16/02/2022)

8. https://habitat-worldmap.org/pays/afrique/senegal/ (Accessed on 20/02/2020)

9. https://www.futura-sciences.com/maison/definitions/maison-eco-construction-10652/ (Accessed on 04/02/2020)

10. https://fr.wikipedia.org/wiki/%C3%89coconstruction (Accessed on 04/02/2020)

11. https://www.planete-senegal.com/senegal/habitat_traditionnel_senegal.php" (Accessed on 20/02/2020)

12. https://afex.squarespace.com/grand-prix/2015/11/7/grand-prix-afex-2012 (Accessed on 06/03/2020)

13. https://www.lepoint.fr/economie/architecture-l-afrique-pourrait-en-finir-avec-le-tout-beton-09-03-2017-2110510_28.php (Accessed on 20/02/2020)

14. https://www.mediaterre.org/habitat/actu,20150525155743,11.html (Accessed on 06/03/2020)

15. http://xalimasn.com/habitat-et-emergence-au-senegal/ (Accessed on 20/02/2020)

16. https://www.habiter-autrement.org/37-ecovillage-afrique/contributions-37/30-Senegal-villa-ecologique-Assie-Gaye.pdf

17. http://www.uadb.edu.sn/index.php/component/k2/item/109-le-nouveau-batiment-de-l-universite-alioune-diop-laureat-du-prix-aga-khan-d-architecture-2019 (Accessed on 06/03/2020)

18. https://www.google.com/search?q=batiment+bioclimatique+Bambey (Accessed on 06/03/2020)

19. https://www.construction21.org/france/articles/fr/22-l-ecopavillon-de-diamniado.html (Accessed on 06/03/2020)

20. https://www.lavoutenubienne.org/senegal/ (Accessed on 01/05/2020)

21. http://bamada.net/lutte-contre-le-changement-climatique-la-voute-nubienne-une-solution-innovante (Accessed on 01/05/2020)

22. http://www.rfi.fr/fr/science/20151124-senegal-construction-voute-nubienne-materiau-retour-terre (Accessed on 07/05/2020)

23. www.au-senegal.com/la-voute-nubienne-retour-vers-les-materiaux-locaux,11876.html (Accessed on 01/05/2020)

24. http://www.fasozine.com/actualite/environnement/6744-la-voute-nubienne-une-technique-a-la-fois-ancestrale-et-innovante-adaptee-aux-changements-climatiques.html (Accessed on 01/05/2020)

25. https://www.construction21.org/france/articles/fr/le-programme-de-developpement-de-l-association-la-voute-nubienne-techniques-et-formations-inclusives-portees-par-une-methodologie-de-diffusion-innovante.html (Accessed on 01/05/2020)

26. https://www.senegal-export.com/bloc-de-terre-comprime-innovation-btp-au-village-pilote.html (Accessed on 17/05/2020)

27. https://fr.wikipedia.org/wiki/Brique_de_terre_compress%C3%A9e (Accessed on 17/05/2020)

28. https://www.oskam-vf.com/blocs_%20de_terre_comprimee.html (Accessed on 17/05/2020)

29. https://www.batiproduits.com/batiwiki/bloc-de-terre-comprim/definition/605783C3-E7FE-4E8D-998C-7C6F8D9D1C32/ (Accessed on 17/05/2020)

30. https://www.senegal-export.com/la-gomme-arabique-un-produit-a-usage-multiple.html (Accessed on 19/11/2021)

31. https://fr.wikipedia.org/wiki/Grand_Prix_AFEX_de_l%E2%80%99architecture_fran%C3%A7aise_dans_le_monde (Accessed on 19/11/2021)

32. https://fr.wikipedia.org/wiki/Diolas (Accessed on 23/11/2021)

33. http://senegalinfo.free.fr/senegal-origine.html (Accessed on 26/02/2022)

34. https://www.construction21.org/france/articles/h/le-programme-de-developpement-de-l-association-la-voute-nubienne-techniques-et-formations-inclusives-portees-par-une-methodologie-de-diffusion-innovante.html (Accessed on 27/02/2022)

35. https://www.dwf.org/fr/francais/qui-sommes-nous (Accessed on 27/02/2022)

36. https://www.lavoutenubienne.org/?lang=fr (Accessed on 27/02/2022)

37.https://www.lavoutenubienne.org/-resultats-et-impacts- (Accessed on 28/02/2022)

38.https://www.lavoutenubienne.org/senegal (Accessed on 28/02/2022)

39.https://search.brave.com/images?q=image+btc+construction (Accessed on 10/07/2022)

40.https://www.green-economy.fr/definition-energie-renouvelable/ (Accessed on 17/02/2020)

41.https://www.contrepoints.org/2019/08/12/351191-les-energies-alternatives-renouvelables-peu-renouvelables-et-tres-alternatives (Accessed on 17/02/2020)

42.https://bfmbusiness.bfmtv.com/entreprise/energies-alternatives-marche-ne-marche-pas-568216.html (Accessed on 18/02/2020)

43.http://www.energie.gouv.sn/economie-denergie/ (Accessed on 07/03/2020)

44.http://www.energie.gouv.sn/mot-de-bienvenue-du-ministre-du-petrole-et-des-energies/ (Accessed on 08/03/2020)

45.https://lenergeek.com/2019/08/14/senegal-energies-renouvelables-macky-sall/ (Accessed on 08/03/2020)

46.http://www.aser.sn/index.php/mot-du-dg/ (Accessed on 08/03/2020)

47.http://www.eaux-forets.sn/?page_id=1751 (Accessed on 12/03/2020)

48.www.lequotidien.sn/inauguration-de-la-centrale-de-taiba-ndiaye-le-senegal-dans-lere-de-leolienne/ (Accessed on 12/03/2020)

49.https://www.jeuneafrique.com/515854/economie/le-senegal-inaugure-la-centrale-solaire-de-ten-merina/ (Accessed on 12/03/2020)

50.https://www.senenews.com/actualites/economie/energie-2/le-podere-ambitionne-de-reduire-a-lhorizon-2017-le-deficit-energetique-dans-lespace-uemoa_94111.html (Accessed on 14/03/2020)

51.https://invention-europe.com/2016/11/11/abdoulaye-toure-un-ingenieur-dans-le-champ-de-captage-de-lenergie-solaire/ (Accessed on 14/03/2020

52. http://www.infos15.com/les-energies-renouvelables-et-alternatives-au-senegal-3eme-partie.html (Accessed on 07/03/2020)

53. http://www.taibaeolien.com/sn/a-propos/ (Accessed on 09/03/2020)

54. https://www.agenceecofin.com/eolien/2502-74178-senegal-la-centrale-eolienne-de-taiba-n-diaye-officiellement-inauguree (Accessed on 14/03/2020)

55. https://www.pe-omvg.org/leprojet%20Consulté%20le%2019.03.2020" (Accessed on 19/03/2020)

56. http://www.nebeday.org/p/charbon-de-paille.html%20consulté%20le%2027.05.2020 (Accessed on 27/05/2020)

57. https://wiki.lowtechlab.org/wiki/Bio_Charbon%20consulté%20le%2027.05.2020 (Accessed on 27/05/2020)

58. http://www.grandirausenegal.fr/biochar-un-dossier-ecologique-qui-permettra-a-des-familles-denvoyer-leurs-enfants-a-lecole/ (Accessed on 27/05/2020)

59. https://www.airex-energy.com/fr/biocharbon (Accessed on 27/05/2020)

60. https://www.gret.org/2018/06/biocharbon-typha-premiere-etape-vers-production-industrielle (Accessed on 27/05/2020)

61. https://www.pthorticulture.com/fr/zone-du-savoir/le-biocharbon-ingredient-potentiel-pour-substrats-de-croissance/ (Accessed on 27/05/2020)

62. https://www.laterre.ca/utiliterre/nouveautes/biocharbon-percee-quebec (Accessed on 27/05/2020)

63. https://www.refertil.info/sme/fr/quest-ce-que-cest-est-le-biocharbon-et-comment-lon-produit (Accessed on 27/05/2020)

64. https://www.koldanews.com/2019/09/16/vels-company-la-start-up-senegalaise-qui-propose-du-bio-charbon-a1037684 (Accessed on 27/05/2020)

65.https://www.actu-environnement.com/ae/dictionnaire_environnement/definition/parc_eolien.php4 (Accessed on 11/03/2022)

66.https://www.topito.com/top-parcs-eoliens-monde (Accessed on 11/03/2022)

67.https://www.les-energies-renouvelables.eu/conseils/eolienne/principe-fonctionnement-eolienne/ (Accessed on 24/04/2022)

68.https://www.engie-green.fr/wp-content/uploads/2020/12/fiche-fonctionnement-bd4.pdf (Accessed on 24/04/2022)

69. http://www.nebeday.org/p/contact.html (Accessed on 23/07/2022)

70.https://www.ideassonline.org/public/pdf/BrochureBiogaz-PNBS.pdf

71.https://lepetitjournal.com/dakar/actualites/la-recherche-au-service-des-agriculteurs-senegalais-242123 (Accessed on 01/11/2019)

72.https://business.freefrenchartricles.com/index.php/2018/05/10/prdication-future-pour-le-march-des-biofertilisants.html (Accessed on 01/11/2019)

73.http://www.elephant-vert.com/le-groupe/qui-sommes-nous/elephant-vert-senegal/ (Accessed on 19/10/2019)

74.https://www.antenna.ch/fr/activites/agriculture/ (Accessed on 19/10/2019)

75.http://www.fao.org/3/x6915f/x6915f03.htm (Accessed on 21/10/2019)

76.http://www.elephant-vert.com/actualites-medias/actualites/la-biotechnologie-au-service-dune-agriculture-durable 5Accessed on 01/11/2019)

77.https://www.seneplus.com/article/le-bio-fertilisant-augmente-la-production-et-restaure-les-sols (Accessed on en 2019)

78.http://www.fao.org/3/x6915f/x6915f03.htm (Accessed on 19/10/2019)

79.https://tribuneoccitanie.com/part-de-marche-mondiale-croissante-des-biofertilisants-segments-principaux-moteurs-et-analyse-de-la-croissance-jusquen-2019-2025/ (Accessed on 30/11/2019)

80.https:.//www.memoireonline.com/01/16/9399/Installation-d-une-unite-de-production-de-biofertilisants.html (Accessed on 27/10/2019)

81.http://www.rfi.fr/emission/20180331-biofertilisants-plantes-captent-ressources-sol (Accessed on 2019)

82.https://www.m-habitat.fr/amenagement-de-jardin/jardin-potager/comment-creer-un-micro-jardin-4204_A (Accessed on 16/12/2019)

83.https://www.nouvelobs.com/rue89/rue89-planete/20150725.RUE9963/exploit-dans-son-micro-jardin-joseph-produit-300-kilos-de-legumes.html (Accessed on 16/12/2019)

84.https://amenagement-de-jardin.ooreka.fr/astuce/voir/287309/principe-et-caracteristiques-des-micro-jardins (Accessed on 16/12/2019)

85.http://www.fairebien.com/agriculture-durable/ (Accessed on 04/02/2020)

86.https://www.geo.fr/environnement/agriculture-durable-definition-principes-et-enjeux-193859 (Accessed on 04/02/2020)

87.https://agridurable.wordpress.com/agriculture-durable/ (Accessed on 04/02/2020)

88.http://www.fao.org/sustainable-development-goals/overview/fao-and-post-2015/sustainable-agriculture/fr/ (Accessed on 04/02/2020)

89.https://fr.allafrica.com/stories/201804170447.html (Accessed on 24/03/2020)

90.https://senego.com/bilan-du-pracas-lagriculture-senegalaise-sest-diversifiee-bonifiee-et-fortifiee_672834.html (Accessed on 24/03/2020)

91.http://www.maer.gouv.sn/directions-et-services-rattaches/ (Accessed on 02/04/2020)

92.https://afrique-ouest.cirad.fr/actualites/vers-un-plaidoyer-national-pour-l-agroecologie-au-senegal (Accessed on 29/03/2020)

93.https://www.upadi-agri.org/federation-des-producteurs-maraichers-des-niayes-fpmn/ (Accessed on 04/04/2020)

94. http://www.acra.it/index.php?option=com_content (Accessed on 04/04/2020)

95. https://www.endapronat.org/ (Accessed on 04/04/2020)

96. http://www.iedafrique.org/-IED-Afrique-.html (Accessed on 04/04/2020)

97. http://www.iedafrique.org/30-Promotion-de-l-agroecologie-au.html (Accessed on 04/04/2020)

98. http://agrecolafrique.org/qui-sommes-nous/presentation/ (Accessed on 04/04/2020)

99. http://www.agrisud.org/nous-connaitre/manifeste-agrisud/ (Accessed on 04/04/2020)

100. http://www.upadi-agri.org/agricord (Accessed on 04/04/2020)

101. https://www.actioncontrelafaim.org/nous-decouvrir/ (Accessed on 04/04/2020)

102. http://fenab.org/index.php/presentation-de-la-feneb/ (Accessed on 06/04/2020)

103. https://www.lequotidien.sn/agriculture-durable-a-kaolack-les-freins-passes-au-peigne-fin/ (Accessed on 04/04/2020)

104. http://yeesalhub.org/agroecologie-au-senegal/ (Accessed on 04/04/2020)

105. https://agriculture.gouv.fr/lagriculture-biologique-1 (Accessed on 16/04/2020)

106. https://agriculture.gouv.fr/quest-ce-que-lagroecologie (Accessed on 16/04/2020)

107. http://www.biotechservices-senegal.com/about-us/ (Accessed on 03/08/2022)

108. https://journals.openedition.org/vertigo/docannexe/image/20120/img-1.jpg (Accessed on 05/07/2023)

Media library

1. https://www.youtube.com/watch?v=6nm5IJuz1mU[COP23] Mbacké NIANG - Architecte de la ville de Dakar, chercheur et enseignant, 6mn 26s

2. https://www.ffem.fr/fr/carte-des-projets/tyccao-typha-combustible-construction-afrique-de-louest Reportage sur le biocharbon sur TV5

3. https://www.youtube.com/watch?v=rdYw8LbLO Reportage sur le BTC sur TV5

4. https://www.youtube.com/watch?v=oFoKzNKN--0 TV5 monde, Sénégal : un parc éolien provoque la colère des habitants de Taïba Ndiaye à 100km de Dakar, 1mn 38 s

5. https://www.youtube.com/watch?v=LOD6b-hucWg RTI, 2020 : Made In Africa : Taïba N'Diaye, l'ambitieux parc éolien du Sénégal pour le défi du vert. 11mn, 49s

6. https://www.youtube.com/watch?v=pJ8fvPNpBM8 Mamadou Sakho, 2021. Sénégal Vert Horizon 2035 – Documentaire, 34 mn, 50 s

7. https://www.youtube.com/watch?v=sazvYUXDFbo Sam Médias Groupe, Enquete exclusive- Découverte du Parc éolien de Taïba Ndiaye au Sénégal, 12 mn 33s

8. https://www.youtube.com/watch?v=WaWCYgIpxvc, Atelier-BK Architecture, 2021. A Dakar au Sénégal : Une villa moderne en briques de terre comprimée (BTC), 42 s.

9. https://www.youtube.com/watch?v=rdYw8LbLO TV5 Monde, 2021. Sénégal : la construction en brique de terre peine à trouver sa place, 2 m 20 s.

10. https://www.youtube.com/watch?v=wbqa5NreKZs Habiteco 127, 2021. Les challenges de construire avec des briques en terre cuite en Afrique, au Sénégal dans le Baol, 5 m 32 s.

11. https://www.youtube.com/watch?v=T5f1jXycwuU MEDI1TV, 2020. Sénégal: le charbon vert pour promouvoir l'entrepreneuriat écologique, 26 m 05 s

www.ingramcontent.com/pod-product-compliance
Lightning Source LLC
LaVergne TN
LVHW022041200926
848481LV00003B/736
* 9 7 9 8 8 9 9 6 6 2 3 9 3 *